F. W Merchant

High School Physical Science

Part Two

F. W Merchant

High School Physical Science
Part Two

ISBN/EAN: 9783741187599

Manufactured in Europe, USA, Canada, Australia, Japa

Cover: Foto ©berggeist007 / pixelio.de

Manufactured and distributed by brebook publishing software
(www.brebook.com)

F. W Merchant

High School Physical Science

HIGH SCHOOL

PHYSICAL SCIENCE

PART II.

BY

F. W. MERCHANT, M.A.,

Collegiate Institute, London.

Authorized by the Department of Education for Ontario.

TORONTO:

THE COPP, CLARK COMPANY, LIMITED.

1896.

CONTENTS.

[iii]

CHAPTER XXIV.

CHAPTER XXV.

CHAPTER XXVI.

CHAPTER XXVII.

CHAPTER XXVIII.

CHAPTER XXIX.

CHAPTER XXX.

PHYSICAL SCIENCE

PART II.

CHAPTER I,

VELOCITY.

The subjects of velocity and acceleration are treated experimentally in Chapter iii. of Part I. We shall now give general statements of the principles there discussed and a number of exercises containing problems in illustration of them.

1. Position.

Position is **relative,** not **absolute.** The position of a point P is determined when its **distance** and its **direction** from some other point, taken as a point of reference, are known.

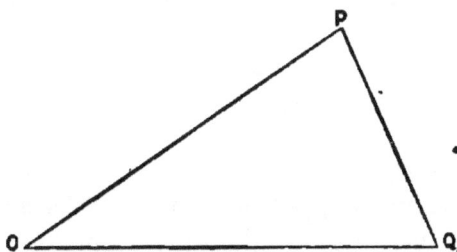

FIG. 1.

The line OP (Fig. 1) represents by its length and its direction the position of the point P with reference to the point O. Similarly, QP represents the position of the point P with reference to the point Q. That is, P has

[1]

one position with reference to O, and another with reference to Q. In the same way, Q has one position, represented by the line OQ, with reference to O, and another, represented by the line PQ, with reference to P.

2. Motion.

A point is said to be in motion when its position is being changed continuously.

Motion, like position, is **relative.** A point is moving relatively to any point of reference when its position with respect to that point is changing continuously.

If the position of a point P with respect to a point of reference remains unchanged for a given time, P is said to be at rest with respect to this point of reference during that time; but, during the same time, the point P may be in motion with respect to another point of reference. A seat in a railway carriage in motion is at rest with respect to another seat in the same carriage, but in motion with respect to any object which has not the same motion as the carriage.

Two points are at rest relatively to each other when their motions are identical.

3. Displacement.

The change in the position of a point for any specified interval of time is called the **displacement** of the point for that interval.

FIG. 2.

If the point P is in motion relatively to the point A in the line AD (Fig. 2), and if at any instant it is at B, and at a subsequent instant at C, BC is the displacement of the point for the interval of time between these instants.

A displacement is determined when its magnitude and its direction are known.

The magnitude of the displacement is measured by the number of units of length contained in the line joining the two positions of the point.

If the point P has successive displacements BC, CD (Fig. 2) in the same direction, the total displacement equals

$$BC + CD,$$

the sum of these displacements; but if the point has displacement, BC, CD, DE, EF (Fig. 2), some in one direction and some in the opposite direction, and the displacements in opposite directions are given opposite signs, the total displacement in the positive direction equals

$$BC + CD - DE - EF,$$

the algebraic sum of the displacements.

4. Velocity.

The amount of time required for a given displacement of a point depends upon the rapidity of the movement of the point. The time-rate at which the displacement takes place is, therefore, an important quantity.

The time-rate of displacement is called velocity.

5. Uniform Velocity.

A point is said to be moving with a **uniform velocity** when it has equal displacements in equal intervals of time, however short these intervals may be.

6. Measure of Velocity.

A velocity, like a displacement, is determined when its magnitude and its direction are known. The magnitude of a velocity is measured by the number of times it contains some definite velocity, assumed as a unit. **The unit velocity is the velocity of a point which, moving uniformly, has a unit displacement in a unit of time.** The unit is a derived unit, because it is determined by the unit of length and the unit of time. For example, when the unit of length is the centimetre, and the unit of time the second, the unit of velocity is the centimetre per second.

With this system of units the measure of the velocity of a point moving uniformly is 1, when it has a displacement of 1 cm. in 1 second; 2, when it has a displacement of 2 cm. in 1 second; 3, when it has a displacement of 3 cm. in one second, and so on. If its displacement is 3 cm. in 2 seconds, in one second it is $\frac{3}{2}$ cm., and the measure of the velocity of the point is $\frac{3}{2}$.

In general terms, the measure of the velocity of a point moving uniformly, equals the measure of its displacement in a unit of time.

Thus, if s is the number of units of length in the displacement, and t the number of units of time in the interval, the measure of the displacement in a unit of time is $\frac{s}{t}$. Then, if v is the measure of the velocity,

$$v = \frac{s}{t}$$

or $s = t\,v.$

That is, **the measure of the velocity** of a point moving

uniformly during a given interval of time, **is obtained by dividing the measure of the displacement during that interval by the measure of the interval,**

or

the measure of the displacement equals the measure of the velocity multiplied by the measure of the interval.

EXERCISE I.

1. Express in miles per hour a velocity of (1) 40 feet per second, (2) 100 yards per minute.

2. A point moves at the rate of 50 miles in $1\frac{1}{2}$ hours. What is its velocity in feet per second ?

3. Find the ratio of velocities of (1) 60 miles per hour and 44 feet per second, (2) 5 miles per 6 minutes and 10 feet per $\frac{1}{4}$ second.

4. One body moves over 30 yards in 7 minutes, and another over 12 feet in 5 seconds. If their velocities are uniform, compare them.

v   5. How many times does the velocity 176 yards per hour contain the velocity 2 miles per minute ?

v   6. A velocity of 20 miles per hour is v times a velocity of 30 feet per second. What is v ?

7. How many times does a velocity of 120 metres per minute contain a velocity of 20 cm. per sec. ?

8. What is the measure of a velocity of 400 cm. per second when the unit velocity is (1) a centimetre per second, (2) a centimetre per 2 seconds, (3) a centimetre per $\frac{1}{2}$ second, (4) 2 centimetres per second, (5) $\frac{1}{2}$ centimetre per second?

9. What is the measure of a velocity of 120 metres per minute when 20 centimetres per second is the unit of velocity ?

10. When 3 metres per 7 seconds is the unit of velocity, what is the measure of a velocity of 126 cm. per 3 seconds ?

11. What is the measure of a velocity of 40 miles per hour when the unit velocity is (1) a foot per minute, (2) a foot per second, (3) a foot per 2 minutes, (4) 2 feet per second ?

12. A point has a uniform velocity of 8 feet per second. What is its displacement in 11 hours?

13. A point is moving with a uniform velocity of 20 cm. per second. What is its displacement in metres in 10 hours?

14. A point moves uniformly in a straight line at the rate of a feet per second. What is its displacement in miles in b hours?

15. A point is moving uniformly at the rate of c cm. in s seconds. How far does it go in h hours?

16. The velocity of a train is 15 miles per hour. Find (1) how many minutes it will take to go 50 yards, (2) how many seconds it will take to go 25 feet.

17. The velocity of a point is a feet per b seconds. How long does it take it to go c miles?

------

7. Velocity at an Instant.

An instant of time has no duration and therefore, while a point may be said to have a velocity at any given instant, an interval of time is necessary to produce a finite displacement, however small. For example, a falling body may be said to have a velocity of 10 feet per second at the instant it comes in contact with the earth, but at that instant its displacement is nothing. What is meant is, that if the body were to continue to move for one second with the velocity which it has at the instant it strikes the ground, it would have a displacement of 10 feet.

8. Average Velocity.

When a point is moving with a variable velocity, its **average velocity** for any given instant of time equals the uniform velocity of another point which has an equal total displacement in the interval.

Hence the measure of the average velocity during a given interval is obtained by dividing the measure of the distance traversed during that interval by the measure of the interval.

Thus, if s is the number of units in the total displacement, t the number of units of time in the interval, and v the measure of the average velocity, $v = \dfrac{s}{t}$.

If the velocity is increasing or decreasing uniformly, the average velocity for any given interval is the velocity at the middle instant of the interval. This equals half the sum of the velocities at the initial and the final instants of the interval.

If u is the measure of the velocity at the initial, and v the measure of the velocity at the final instant,

$$\text{the average velocity} = \frac{u+v}{2}$$

and if t is the measure of the interval, and s the measure of the displacement,

$$s = \left[\frac{u+v}{2} \right] t.$$

9. Measure of Variable Velocity.

If a point is moving with a varying velocity its actual velocity at a given instant may be defined as the average velocity during an infinitely short interval containing that instant.

A variable velocity may, therefore, be approximately measured by determining the average velocity for an interval containing the instant. It is manifest that the accuracy of the determination will depend upon the length of the interval. The shorter the interval the more accurate is the result. For example, the speed of a trolley car at a certain instant cannot be accurately determined by taking the average velocity between two stopping places; but, if the space traversed by the car in a very short interval containing the instant is observed, and the average velocity for the instant calculated, the result will be approximately the velocity of the car at the instant.

EXERCISE II.

1. A point has displacements of 3 feet, 4 feet, 5 feet, and 6 feet in four consecutive seconds. What is its average velocity?

2. A point has displacements of 9 cm., 10 cm., 11 cm., and 12 cm. in four consecutive seconds. Find its average velocity (1) for the four seconds, (2) for the first three seconds, (3) for the last three seconds.

3. A point is displaced 5 cm., 3 cm., 1 cm., -1 cm., -3 cm. in five consecutive seconds. What is its average velocity (1) for the five seconds, (2) for the first 3 seconds, (3) for the last three seconds, (4) for the middle three seconds?

4. A point has displacements of 2 feet, 6 feet, 10 feet, 14 feet, and 18 feet in five consecutive seconds. Show that the average velocities for the five seconds, the middle second, and the three middle seconds, are all equal.

5. A point moving with a uniformly increasing velocity has a velocity of 10 feet per second at the beginning of a certain interval of 5 seconds, and a velocity of 20 feet per second at the end of the interval. Find (1) its average velocity, (2) its displacement, during the interval.

6. A point moving with a uniformly decreasing velocity has a velocity of 60 cm. per second at the beginning of a certain interval of 10 seconds, and a velocity of 10 cm. per second at the end of the interval. Find (1) the velocity at the middle instant of the interval, (2) the displacement during the interval.

7. A point has a velocity of 10 cm. per second at the beginning, and 12 cm. per second at the end of a certain second. If its velocity is increasing uniformly, find (1) its velocity at the end of three seconds more, (2) its average velocity for the four seconds, (3) its displacement for the four seconds.

✓ **8.** In one hour the velocity of a point decreases uniformly from 200 feet per second to 100 feet per second. What is the velocity at the end of each quarter of an hour, and what is the total displacement during the hour?

✓ **9.** At 9 A.M. a point is moving with a velocity of 40 cm. per second, and at 1 P.M. it has a velocity of 120 cm. per second. If it moves with a uniformly increasing velocity, what velocity will it have at 11.30 A.M.?

10. At 1 P.M. a point has a velocity of 60 feet per second, and in one minute its velocity increases to 65 feet per second. If the velocity continues to increase uniformly at the same rate, find (1) its velocity at 2.15 P.M., (2) how far it goes between 1.15 and 1.20 P.M., (3) how far it goes between 2.10 and 2.15 P.M.

11. A point is moving with a uniformly decreasing motion. At the beginning of a certain second its velocity is 20 feet per second, and at the end of the same second its velocity is 18 feet per second. When will it come to rest? How far will it go during the second before it comes to rest?

12. A point is moving at a given instant with a velocity of 8 feet per second. At the end of 5 seconds its velocity is 19 feet per second. When will it have a velocity of (1) 16.8 feet per second, (2) 30 feet per second?

13. A point, starting from rest, has its velocity increased uniformly 5 feet per second each second. Find (1) the velocity at the end of 10 seconds, (2) the displacement of the point during that time.

14. A point which has a velocity of 80 centimetres per second has its velocity uniformly decreased by 4 cm. per second each second. How long before it will come to rest, and how far will it move during the time?

15. The velocity of a point increases uniformly in 10 seconds from 150 cm. per second to 200 cm. per second. Find (1) the rate of increase in velocity, (2) the displacement during the 10 seconds.

16. A train moving with a velocity of 60 kilometres per hour is pulled up with a uniformly decreasing velocity in 30 seconds. What is the decrease in velocity during each second in centimetres per second?

CHAPTER II.

1. Uniform Acceleration.

If the motion of a point is changing, the point is said to be accelerated positively or negatively, according as its velocity is increasing or diminishing.

Rate of change of velocity is called Acceleration.

The acceleration is **uniform** when equal changes of velocity take place in equal intervals of time, however short these intervals may be.

2. Measure of Uniform Acceleration.

Since acceleration is the rate of change of velocity, or the change of velocity in a unit of time, acceleration is measured by a unit of acceleration derived from the unit of velocity and the unit of time.

The unit acceleration is the acceleration of a point, the motion of which is such that its velocity is increased, or diminished, by the unit of velocity in each unit of time.

For example, if the centimetre is taken as the unit of length and the second as the unit of time, the **unit velocity is the centimetre per second,** and the **unit acceleration the centimetre per second per second.** An acceleration of 1 cm. per second per second is a change in velocity, during one second, of one centimetre per second; an acceleration of 2 cm. per second per second is

[11]

a change in velocity, during one second, of two centimetres per second; and an acceleration of a cm. per second per second, a change in velocity, during one second, of a centimetres per second.

If v is the measure of the change in velocity of a point in an interval of time, the measure of which is t, $\frac{v}{t}$ is the measure of the change of velocity in a unit of time. Then, if a is the measure of the acceleration, or rate of change of velocity,

$$a = \frac{v}{t}$$

or $\qquad \mathbf{v} = a\ \mathbf{t}.$

That is, the **measure of the uniform acceleration** of a point during a given interval of time is **obtained by dividing the measure of the change of velocity during that interval by the measure of the interval,**

or

the measure of the change in velocity equals the measure of the acceleration multiplied by the measure of the interval.

EXERCISE III.

1. A point is moving with a uniform acceleration of 10 units of velocity per second. (1) What velocity will it acquire in a minute? (2) What will be the acceleration in units of velocity per minute?

2. A point is moving with a uniform acceleration of 10 feet per second per second. (1) What is the total change in velocity in a minute? (2) What is the measure of the acceleration in feet per second per minute?

3. What velocity will a body acquire in half an hour if the acceleration is (1) 10 centimetres per second per minute, (2) 10 centimetres per second per second ?

4. A point is travelling with an acceleration of 12 feet per second per hour. Find (1) what will be its change in velocity in a minute, (2) the measure of its acceleration in feet per second per second ?

5. A train acquires a velocity of 30 feet per second in one hour. If its velocity is uniformly accelerated, find (1) the velocity which it will acquire in one minute, (2) the measure of its acceleration in feet per second per second.

✓ 6. A point is travelling with an acceleration of 12 feet per second per hour. How long will it take it to acquire a velocity of 2 feet per second ?   10 min

∨ 7. A train, moving with a uniform acceleration, acquires a velocity of 75 feet per second in a quarter of a minute. How long will it take it to acquire a velocity of 100 yards per minute ?

8. A point, moving with a uniform acceleration, acquires a velocity of 60 feet per second in 10 minutes. What is the measure of its acceleration in (1) feet per second per minute, (2) yards per second per minute, (3) feet per second per second, (4) yards per second per second ?

9. A point, travelling with a uniform acceleration, has its velocity increased 50 metres per second each minute. What is the measure of the acceleration in (1) metres per second per minute, (2) centimetres per second per minute, (3) metres per second per second, (4) centimetres per second per second ?

10. A train, moving with a uniform acceleration, acquires an additional velocity of 60 feet per second each minute. Find (1) the measure of its acceleration in feet per second per second, (2) the measure of the velocity it acquires each minute in feet per minute, (3) the measure of the acceleration in feet per minute per minute, (4) the measure of the acceleration in feet per minute per second.

11. A point is moving with a uniform acceleration and acquires an additional velocity of 20 cm. per second each second. Find the measure of the acceleration in (1) centimetres per second per

minute, (2) centimetres per minute per minute, (3) metres per minute per minute, (4) metres per minute per second, (5) metres per second per second.

12. What is the measure of an acceleration of 30 feet per second per second when the units of displacement and of time are respectively (1) the foot and the second, (2) the foot and the minute?

13. The measure of the acceleration of a falling body is 32 when the foot is the unit space and the second the unit of velocity. What is the measure of this acceleration when the yard is the unit of space and the minute is unit of time?

14. The acceleration due to gravity of a body is 980 centimetres per second per second. What is the measure of this acceleration when the metre is the unit of displacement and the minute the unit of time?

15. Compare an acceleration of 120 feet per second per minute with an acceleration of 20 yards per minute per second.

16. How many times does an acceleration of 5 feet per second per second contain an acceleration of 18 feet per minute per minute?

17. What is the measure of an acceleration of 5 feet per second per second when 18 feet per minute per minute is the unit of acceleration?

18. Compare an acceleration of which the measure is 150 when the yard is the unit displacement and the minute the unit of time with an acceleration of which the measure is .25 when the foot is the unit of displacement and the second the unit of time.

19. What is the measure of an acceleration of 150 yards per minute per minute when an acceleration of .25 feet per second per second is the unit acceleration?

We have in the preceding exercises given a series of questions which illustrate the relations which exist among the quantities involved in problems relating to uniformly accelerated velocities. We shall now derive equations which state in a general way these relations.

Let a be the measure of the acceleration of a point moving with a uniform acceleration.

t, the measure of any interval of time in its motion.

s, the measure of the displacement for the interval.

u, the measure of the velocity of the point at the beginning of the interval.

v, the measure of the velocity of the point at the end of the interval.

Then,

$v - u =$ the change in velocity in t units of time,

$\dfrac{v - u}{t} =$ the change in velocity in one unit of time,

$=$ the rate of change in velocity,

$= a$,

or, $\qquad v = u + a\,t \quad \ldots \ldots$ (i.).

If the point starts from rest, $u = o$ and $v = a\,t$.

Therefore, if the acceleration remains constant,

$$v \propto t.$$

Again, since the velocity increases or decreases uniformly,

the average velocity $=$ the velocity at the middle instant of the interval,

$=$ half the sum of the initial and the final velocities,

$= \dfrac{u + v}{2}$

But the displacement = the average velocity × time (Art. 8, page 7). Therefore,

$$s = \left(\frac{u + v}{2}\right) t \quad . \quad . \quad . \quad . \quad \text{(ii.).}$$

Substituting the value for v given in (i.) for v in (ii.), we have,

$$s = \frac{u + u + at}{2} \times t$$

or, $$s = ut + \tfrac{1}{2}at^2 \quad . \quad . \quad . \quad . \quad \text{(iii.).}$$

It should be carefully noted that when the initial velocity is in the opposite direction to the acceleration, that is when u and a are of opposite signs, the value of s given in this equation does not always indicate the whole distance travelled by the point during the interval of time, but simply the distance between its positions at the beginning and at the end of the interval.

If the point starts from rest, $u = o$, and

$$s = \tfrac{1}{2}at^2.$$

Therefore, if the acceleration is constant,

$$s \propto t^2.$$

From (i.) $$v = u + at,$$

or, $$t = \frac{v - u}{a}$$

From (ii.) $$s = \left(\frac{v + u}{2}\right) t$$

Substituting the value for t given in (i.) for t in (ii.),

We have

$$s = \frac{v+u}{2} \times \frac{v-u}{a} = \frac{v^2-u^2}{2\,a}$$

or $\qquad \mathbf{v}^2 = \mathbf{u}^2 + 2\,a\,\mathbf{s}.$ (iv.).

If the point starts from rest, $u = o$, and

$$v^2 = 2\,a\,s.$$

Therefore, if the acceleration is constant,

$$s \propto v^2.$$

It will be noticed that each of the equations,

$\mathbf{v} = \mathbf{u} + a\,\mathbf{t} \quad \ldots \ldots \ldots$ (i.).

$\mathbf{s} = \left(\dfrac{\mathbf{u}+\mathbf{v}}{2}\right)\mathbf{t} \quad \ldots \ldots \ldots$ (ii.).

$\mathbf{s} = \mathbf{u}\,\mathbf{t} + \tfrac{1}{2}\,a\,\mathbf{t}^2 \quad \ldots \ldots \ldots$ (iii.).

$\mathbf{v}^2 = \mathbf{u}^2 + 2\,a\,\mathbf{s} \quad \ldots \ldots \ldots$ (iv.).

gives the relation among four of the five quantities $\mathbf{u}$, $\mathbf{v}$, $\mathbf{t}$, a, and $\mathbf{s}$, and that, if any three of these are known, the values of the other two may be derived from the equations.

EXERCISE IV.

1. What is the initial velocity of a point which, moving with a uniform acceleration of 10 centimetres per second per second, acquires in 10 seconds a velocity of 200 centimetres per second?

2. A body, moving at a certain instant with a velocity of 30 miles per hour, is subject to a uniform acceleration in the opposite direction, and comes to rest in 11 seconds. What was the measure of its velocity, in feet per second, 5 seconds before it stopped?

2

3. Find the initial velocity of a point which moves with a uniform acceleration of 20 centimetres per second per second, and acquires a velocity of 15 centimetres per second in 10 seconds. Interpret the result.   •

4. The velocity of a point increases uniformly in 20 seconds' from 100 centimetres per second to 200 centimetres per second. Find (1) the measure of the acceleration in centimetres per second per second, (2) the velocity three seconds after it was 150 centimetres per second, (3) when the body was at rest.

5. A point, which has an acceleration of 32 feet per second per second, is moving with a velocity of 10 feet per second. At the same place and at the same time another point, which has an acceleration of 16 feet per second per second, is moving in the same direction with a velocity of 170 feet per second. Find (1) when the two points will have equal velocities, (2) when the velocity of the second will be double that of the first.

6. A body, moving with a velocity of 5 centimetres per second, has a constant acceleration of 10 centimetres per second per second in the direction of its motion. Find (1) how far it will go in 10 seconds, (2) how long it will take it to go 10 centimetres.

7. A body starts with a velocity of 15 centimetres per second, and has a constant acceleration of 10 centimetres per second per second in the opposite direction. When and where will it come to rest?   ·

8. A body, starting from rest, moves with a uniform acceleration of 20 feet per second per second. Find (1) how far the body goes in 4 seconds, (2) how far it goes in five seconds, (3) how far it goes in the fifth second.

9. A body starts with a velocity of 6 feet per second and has a uniform acceleration of 3 feet per second per second in the direction of its motion. At the end of four seconds the acceleration ceases. How far does the body move in 10 seconds from the beginning of its motion ?

10. With what uniform acceleration does a point, starting from rest, describe 640 feet in 8 seconds ?

✓11. A point, starting from rest and moving with a uniform acceleration, has a displacement of 66 feet in the 6th second. What is the measure of the acceleration in feet per second per second, and what is its displacement in the 7th second ?

✎ 12. A body, moving with a uniform acceleration, has displacements in the 4th and the 6th seconds from starting of 38 feet and 54 feet respectively. Find (1) the acceleration, (2) the initial velocity, (3) the displacement in the 7th second.

13. A particle, moving with a uniform acceleration, has a displacement of 30 centimetres in the first 2 seconds of its motion. The acceleration then ceases, and the displacement for the next 2 seconds is 20 centimetres. Find (1) the initial velocity, (2) the acceleration.

14. A train, having a velocity of 20 feet per second, attains a velocity of 30 miles per hour in passing over 128 feet. If the train is moving with a uniform acceleration, what is its acceleration ?

∨ 15. A trolley car, moving at the rate of 24 feet per second, is stopped with a uniformly decreasing motion in a space of 9 feet. What is the acceleration of the car ?

16. A particle starts with a velocity of 23 feet per second, and its velocity is uniformly decreased at the rate of 8 feet per second per second. Find how long it will take it to describe a distance of 30 feet, and how much longer to come to rest.

17. The displacement of a point moving with a uniformly decreasing motion is 100 metres in 10 seconds, and 100 metres in the next 12 seconds. In what time will it be 100 metres further on ?

18. What is the velocity of a particle which, starting from rest and moving with a uniform acceleration of 8 feet per second per second, has traversed 100 feet ? Find also the time required for this displacement.

19. A point which has a negative uniform acceleration of 10 centimetres per second per second is passing a certain point, and at the end of 15 centimetres farther on it has a velocity of 10 centimetres per second. Find (1) its velocity at the point, (2) in how many seconds it will return to this point.

20. A body starts from rest and moves with a uniform acceleration. If its displacement is 90 feet in the fifth second of its motion, find (1) the acceleration, (2) the velocity of the body after 10 seconds.

21. A point, moving with a uniform acceleration, describes in the last second of its motion $\frac{11}{36}$ of the whole distance. If it started from rest, how long was it in motion and through what distance did it move, if it described 4 centimetres in the first second?

22. If a body describes 36 feet while its velocity increases uniformly from 8 feet per second to 10 feet per second, how much further will it go before it attains a velocity of 12 feet per second?

23. A particle moves with a uniform acceleration through 80 feet in 4 seconds, and then comes to rest. Find (1) the initial velocity, (2) the average velocity during each second.

24. A point, moving with a uniform acceleration, attains a velocity of 50 centimetres per second from rest in going 125 centimetres. The acceleration is then changed to an acceleration in the opposite direction, and the point comes to rest in 250 centimetres. Compare the two accelerations.

25. A body, having a uniform acceleration, has a displacement of 27 feet in the fourth second from rest. What was the velocity at the beginning of the fourth second?

26. A particle, having a uniform acceleration, has a displacement of 32.5 cm. in the half second which elapses after the 2nd second of its motion, and a displacement in the 5th second of its motion of 110 cm. Find the initial velocity and the acceleration of the point.

27. The spaces traversed in the 1st, 2nd, 3rd, etc., seconds by a moving body are proportionately 1, 3, 5, etc. Is this consistent with the supposition that it is moving with a uniform acceleration?

28. A particle moving with a uniform acceleration of 40 centimetres per second per second, starts from a given point with a velocity of 40 centimetres per second; and three seconds afterwards another particle starts from the same point in the same direction with a velocity of 30 centimetres per second, and moves with a uniform acceleration of 60 centimetres per second per second. When and where will the second particle overtake the first?

CHAPTER III.

We have learned (Chaps. v. and vi., Part I.) that there is in all bodies at the surface of the earth a **tendency to acceleration**, and that, if unsupported, they begin to move with a **uniform acceleration toward the earth's centre.**

1. Acceleration of a falling body independent of its mass or of the kind of matter of which it is composed.

The tendency to acceleration in a body at the earth's surface is proportional to its mass; hence all bodies, whatever their masses, fall in vacuo with the same acceleration. This may be illustrated by the following experiment ·

Experiment 1.

Place a coin and a feather or small piece of paper in a "guinea and feather tube" (Fig. 3), close the tube, invert it, and observe the motion of the coin and of the feather. Partially exhaust the air, invert the tube, and observe the motion of each. Now exhaust the air as completely as possible, again invert the tube, and observe the motion.

1. How does the exhausting of the air affect the relative velocities of the coin and the feather?

2. Why is the one retarded more by the air than the other?

FIG. 3.

[21]

2. Measure of the Acceleration due to Gravity.

A simple method of determining the acceleration due to gravity is given in Experiment 4, page 70, Part I.

The following method is simple, and if the experiment is performed with care, will be found to give a result which is approximately correct.

Experiment 2.

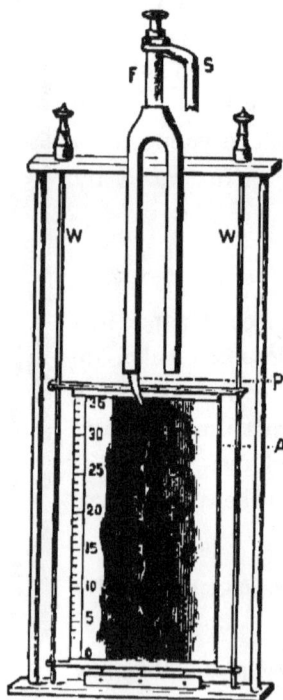

Fig. 4

Arrange apparatus as shown in Fig. 4. The weighted frame A, which has fitted into it a smoked glass upon which is a vertical scale, is arranged to fall down vertical guide wires W, W. A tuning-fork F, which vibrates a known number of times per second, say 100, is placed in a support S and so adjusted that a light aluminium pointer p just touches the smoked glass. The frame is raised to its highest point, and ·is supported there by a thread. The fork is set vibrating, and the thread burned. The frame falls, and the pointer traces a wavy line on the smoked glass. The scale indicates the distance fallen, and the number of waves in the tracing, the number of units of time in the interval. Since s and t are known, the acceleration can be calculated from the equation,

$$s = \tfrac{1}{2} a\, t^2.$$

The apparatus may be placed before the condenser of a porte

lumiere or projection lantern, and the tracing and the scale magnified by projecting them on a screen (Fig. 5).

If a millimetre scale shows a displacement of 31.4 millimetres when the tracing indicates an interval from rest of $\frac{8}{100}$ seconds, what is the acceleration in centimetres per second per second?

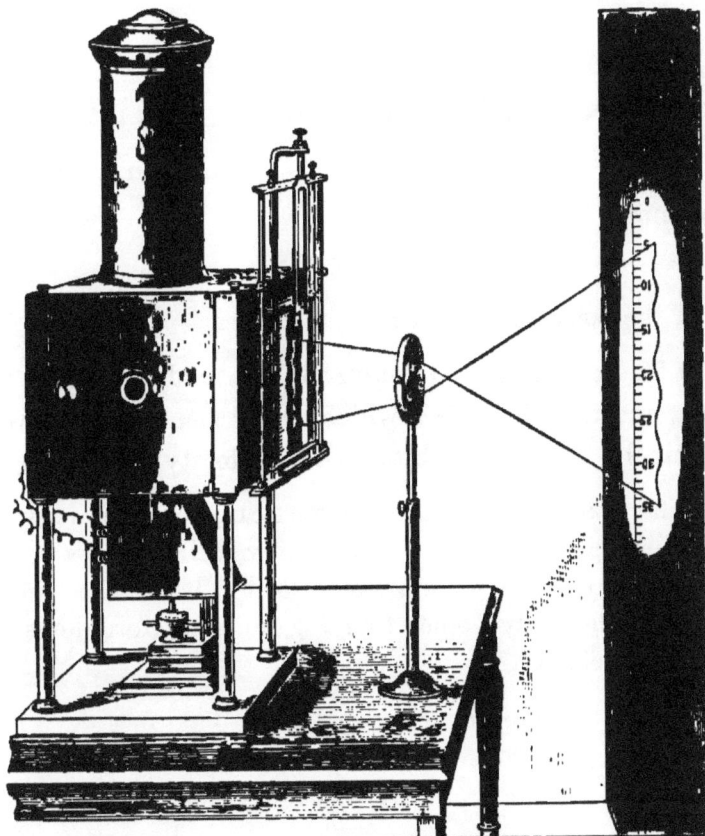

Fig. 5.

3. To find the velocity of a body and the space described by it at the end of an interval of time t, when the body has been thrown vertically downward with an initial velocity of u.

Taking g as the measure of the acceleration due to gravity and substituting in

$$v = u + a\, t \quad . \quad . \quad . \quad \text{(i.), page 17,}$$

we have

$$v = u + g\, t.$$

If the body falls from rest, $u = o$, and $v = g\, t$.

To find the space described, substitute g for a in

$$s = u\, t + \tfrac{1}{2}\, a\, t^2 \quad . \quad . \quad . \quad \text{(iii.), page 17,}$$

therefore

$$s = u\, t + \tfrac{1}{2}\, g\, t^2.$$

If the body falls from rest, $u = o$, and $s = \tfrac{1}{2}\, g\, t^2$.

4. To find the time t required for a body to come to rest, and the distance s, which it will rise when thrown vertically upward with an initial velocity of u.

When the body is thrown vertically upward it will lose in each unit of time g units of velocity, or in t units of time it will lose $g\, t$ units of velocity; but in the t units of time the body comes to rest, and therefore loses u units of velocity.

Therefore $u = g\, t$

or $t = \dfrac{u}{g}$

Again, since the initial velocity is u and terminal velocity $= o$,

The average velocity $= \dfrac{u + o}{2}$ or $\dfrac{u}{2}$

But the distance the body rises

$$= \text{average velocity} \times \text{time},$$

or $\qquad s = \dfrac{u}{2} t = \dfrac{u}{2} \times \dfrac{u}{g} = \dfrac{u^2}{2g}$

5. **To find the displacement s and the velocity v at the end of an interval of time t, when a body is thrown vertically upward with an initial velocity of u.**

When the body is thrown vertically upward it will move with a negative acceleration of g. Substituting $-g$ for a in

$$v = u + a t \quad \ldots \quad \text{(i.), page 17,}$$

we have $\qquad v = u - g t.$

And substituting in

$$s = u t + \tfrac{1}{2} a t^2 \quad \ldots \quad \text{(iii.), page 17,}$$

we have $\qquad s = u t - \tfrac{1}{2} g t^2.$

6. **To find the time t required to reach a height h when a body is thrown vertically upward with a velocity of u. To find also the velocity v at the given height.**

Substituting $-g$ for a, and h for s in

$$s = u t + \tfrac{1}{2} a t^2 \quad \ldots \quad \text{(iii.), page 17,}$$

we have $\qquad h = u t - \tfrac{1}{2} g t^2$

or $\qquad g t^2 - 2 u t + 2 h = 0$

Therefore, $\qquad t = \dfrac{u \pm \sqrt{u^2 - 2 g h}}{g.}$

Both values of t are positive. The lesser gives the time required by the body to reach the given point, and the greater the time required by it to come to rest and fall back to this point.

To find the velocity at the given height, substitute $-g$ for a, and h for s in

$$v^2 = u^2 + 2 \, a \, s. \quad . \quad . \quad \text{(iv.), page 17.}$$

Therefore $v^2 = u^2 - 2 \, g \, h.$

EXERCISE V.

[In solving the following problems the measure of the acceleration due to gravity is to be taken as 32, when the foot is the unit of length and the second is the unit of time; and as 980, when the centimetre is the unit of length and the second the unit of time.]

1. A body drops vertically from rest. What velocity will it have (1) at the end of 5 seconds, (2) when it has fallen 1600 feet?

2. A body is thrown vertically downward with an initial velocity of 100 feet per second. Find what velocity the body will have (1) at the end of 10 seconds, (2) when it has fallen 900 feet.

3. A body is thrown vertically upward with an initial velocity of 4900 centimetres per second. Find its velocity (1) at the end of 3 seconds, (2) when it has risen 117.6 metres.

4. A body falls from rest for 4 seconds. Find the distance fallen (1) in the four seconds, (2) in the fourth second, (3) when it has a velocity of 100 feet per second.

5. A body is thrown vertically downward with an initial velocity of 1470 centimetres per second. Find the distance traversed in the fourth second.

6. A body is thrown vertically upward with an initial velocity of 100 feet per second. Find the height to which it will rise.

7. A body is projected vertically upward with an initial velocity of 160 feet per second. Find the distance traversed (1) in 5 seconds, (2) in the fifth second.

8. A body is thrown vertically upward with an initial velocity of 50 feet per second. What is its height when its velocity is 30 feet per second?

9. A stone is thrown vertically into the shaft of a mine with a velocity of 5.4 metres per second, and reaches the bottom in 4 seconds. What is the depth of the mine?

10. A particle is projected vertically upward with a velocity of 96 feet per second. In what time (1) will its velocity be 48 feet per second, (2) will its displacement be 144 feet? /

11. A body drops vertically from rest. Find (1) when its velocity is 2450 centimetres per second, (2) when the body is 99.225 metres from the point from which it dropped.

12. A stone is projected vertically downward with a velocity of 100 feet per second. Find (1) when its velocity is 292 feet per second, (2) when it is 900 feet from the point of projection.

13. With what velocity must a body be thrown vertically upward (1) that it may rise for 3 seconds, (2) that it may have a velocity of 30 feet per second at the end of the 3rd second, (3) that it may rise 100 feet?

14. With what velocity must a body be thrown vertically downward (1) that it may have a velocity of 100 feet per second at the end of the 2nd second, (2) that it may describe 204 feet in 3 seconds?

15. A body, thrown vertically upward, passes a point 173 feet from the point of projection with a velocity of 50 feet per second. How much further will it go, and what was the velocity with which it was projected?

16. A stone is dropped from a height of 5 metres, and at the same instant another stone is thrown vertically upward, and the stones meet half way. Find the velocity of projection of the latter stone.

17. A particle is projected vertically upward, and it is found that when it is just at a point 39.2 metres from the point of projection it takes 2 seconds to return to the same point. Find (1) the velocity of projection, (2) the whole height ascended.

18. A particle is projected vertically downward, and its displacement in a certain interval of time is 720 feet. In the next

interval of the same length the displacement is 1520 feet. Find the velocity of projection and the measure of the interval.

19. A balloon has been ascending with a uniform velocity for 3 seconds ; and a stone let fall from it reaches the ground in 5 seconds. Find (1) the velocity of the balloon, (2) its height when the stone was let fall.

20. A stone falls from a height, and three seconds after it begins to fall another stone is thrown vertically downward from the point from which the first stone falls, with a velocity of 44.1 metres per second. When and where will the second stone pass the first ?

21. A body projected vertically upward comes to rest at a point 576 feet above the point of projection. When will the body be 144 feet above the point of projection ?

22. A particle is thrown vertically upward and passes a certain point with a velocity of 58.8 metres per second. How long before it will be moving downward at the same rate ?

23. A stone is thrown vertically downward into the shaft of a mine with a velocity of 20 feet per second. If the stone passes through 100 feet in the last second of its fall, find the depth of the mine and the time taken by the stone in reaching the bottom.

24. A stone drops into a mine, and during the last second of its flight falls through ⅝ of the total depth of the mine. What is the depth of the mine and the time taken by the stone in reaching the bottom ?

25. A stone is thrown vertically upward with a velocity of 96 feet per second, and after 4 seconds from the instant of projection another stone is let fall from the same point. In what time will the first stone pass the second ?

26. A body is let fall from the top of a tower 19.6 metres in height, and at the same instant another body is projected vertically upward from the base of the tower with a velocity of 19.6 metres per second. When will the bodies meet ?

27. A balloon is rising with a uniform velocity of 10 feet per second when a stone is dropped from it. If the stone reaches the ground in 3 seconds, find the height of the balloon (1) when the stone was dropped, (2) when the stone reached the ground.

28. A man, descending the shaft of a mine with a uniform velocity of $1\frac{2}{3}$ feet per second, drops a stone which reaches the bottom in 2 seconds. How far did the stone fall?

29. Two particles are simultaneously thrown vertically upward from the same point, the one with an initial velocity of 144 feet per second, and the other with an initial velocity of 202. feet per second. Find the height of the latter when the former reaches the ground.

30. A body after having fallen 3 seconds breaks a pane of glass and thereby loses one-third of its velocity. Find the space through which it falls in 4 seconds.

31. A stone falls freely for 3 seconds, when it passes through a sheet of glass, and in consequence loses one-half of its velocity. If the stone reaches the ground in 2 seconds after passing through the glass, find the height of the glass from the ground.

32. The intensity of gravity at the surface of the planet Jupiter being 2.6 times as great as it is at the surface of the earth, find approximately the time required by a body in falling from a height of 167 feet to the surface of Jupiter.

33. If a heavy body is thrown vertically up to a given height and then falls back to the earth, show that, neglecting the resistance of the air, it passes each point of its path with the same velocity when rising and when falling.

34. A stone is dropped into the shaft of a mine and is heard to strike the bottom in 12.76 seconds. If sound travels at the rate of 1100 feet per second, what is the depth of the mine?

35. A stone dropped into a well reaches the water with a velocity of 80 feet per second, and the sound of its striking the water is heard $2\frac{7}{13}$ seconds after it is let fall. At what velocity did the sound travel?

36. A body has fallen through a feet when another body begins to fall at a point b feet below it. What is the distance which the latter body will fall before it will be passed by the former?

37. If a body is projected vertically upward with a velocity ng (where g is the measure of the acceleration due to gravity), when will its height be ng, and what will then be its velocity?

CHAPTER IV.

1. Force.

The nature of force, its manifestations, and the methods of measuring it are discussed at length in Part I. The student is recommended to review Chapters v. and vii. of that part before proceeding to read this chapter.

2. Representation of a Force.

A force is completely determined when (1) its magnitude, (2) its direction, (3) its point of application are known. These elements of a force may be completely represented by a line. The length of the line may be made to represent the magnitude of the force; the direction of the line, the direction of the force; and an extremity of the line, the point of application of the force.

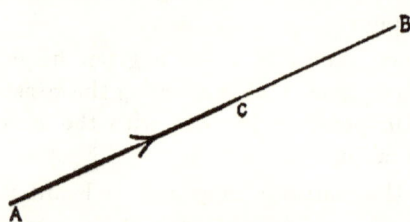

For example, if a line one centimetre in length is taken to represent a gram force, a force of 3 grams, acting at a point denoted by A, and in a direction denoted by AB, will be represented by the line AC (Fig. 6), 3 centimetres in length. An arrowhead is frequently used to indicate the direction in which the force acts.

Fig. 6.

AC represents a force acting in the direction AC, and CA a force acting in the direction CA.

[30]

EXERCISE VI.

1. Taking a line one centimetre in length to represent a gram force, draw a line to represent a force of 12.3 grams acting (1) in a horizontal direction, (2) in a vertical direction, (3) in a direction making an angle of 45° with the horizontal.

2. Taking a line three-quarters of an inch long to represent a pound force, draw a line which represents a force of $5\frac{3}{4}$ pounds acting in a direction making an angle of 60° with the vertical.

A   B   C D

Fig. 7.

3. If AB (Fig. 7) represents a force of 60 grams, what force will be represented by (1) AC, (2) BC, (3) BD, (4) AD, (5) CD?

4. If BC (Fig. 7) represents a force of 24 pounds, what force will be represented by (1) AB, (2) AC, (3) AD, (4) BD, (5) CD?

5. If CD (Fig. 7) represents a force of 3 kilograms, what force will be represented by (1) AB, (2) AC, (3) AD, (4) BC, (5) BD?

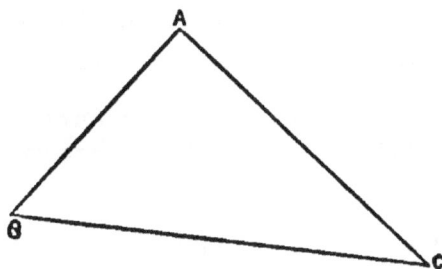

Fig. 8.

6. If 2 cm. in length represents a force of 3 grams, what is the magnitudes of the forces represented by AB, BC, CA, the sides of the triangle ABC? (Fig. 8.)

7. If AD (Fig. 9) represents a force of 2 pounds, what is the magnitudes of the forces represented by AB, AE and ED ?

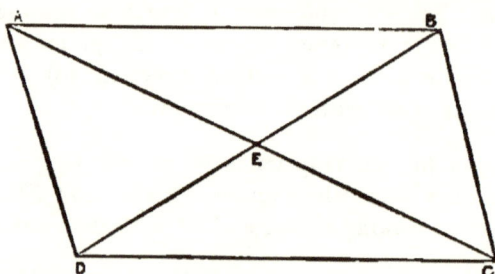

Fig. 9.

3. Resultant and Component Forces.

The single force which represents the combined effect of two or more forces is called the **resultant** of these forces, and the forces themselves are called its **components**. Forces are said to be **compounded** when two or more forces are replaced by a single force equivalent to them; that is, when the resultant is substituted for the components. A force is said to be **resolved** when a single force is replaced by two or more equivalent to it; that is, when the components are substituted for the resultant.

4. To find the resultant of two forces acting at a point when their directions are in the same straight line.

If two forces P and Q, represented by AB and BC, act in the same straight line, it is manifest that their resultant R will be represented by

$$AB + BC \qquad \text{(Fig. 10)}$$

Fig. 10.

that is,

$$R = P + Q$$

when the forces act in the same direction; and by

$$AB - BC \qquad \text{(Fig. 11)}$$

A ——————→————— B ←————— C

FIG. 11.

that is,

$$R = P - Q$$

when the forces act in opposite directions.

5. Equilibrium.

Whenever two or more forces act upon a particle and their individual tendencies to acceleration so counteract each other that no motion results, the forces are said to be in **equilibrium.**

It is evident that if two forces P and Q (Fig. 11) keep a particle in equilibrium, they must be equal in magnitude and act in opposite directions in the same straight line.

If several forces P,Q,R,S,T, in the same plane, acting at a point (Fig. 12), keep a particle in equilibrium, any one of them must equal in magnitude the resultant of all the others, and this force

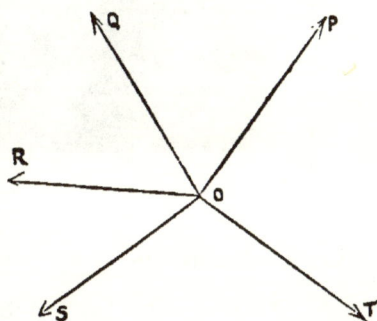

FIG. 12.

and the resultant of the others must act in opposite directions in the same straight line.

3

6. To find the resultant of two forces acting at a point when the directions of the forces are not in the same straight line—Parallelogram of Forces.

Experiment.

Knot the ends of three strings together at O (Fig. 13), and attach the other end of each to a spring balance. Draw the strings tight and attach the balances to hooks in the frame of a black-board or drawing-board, as shown in the figure.

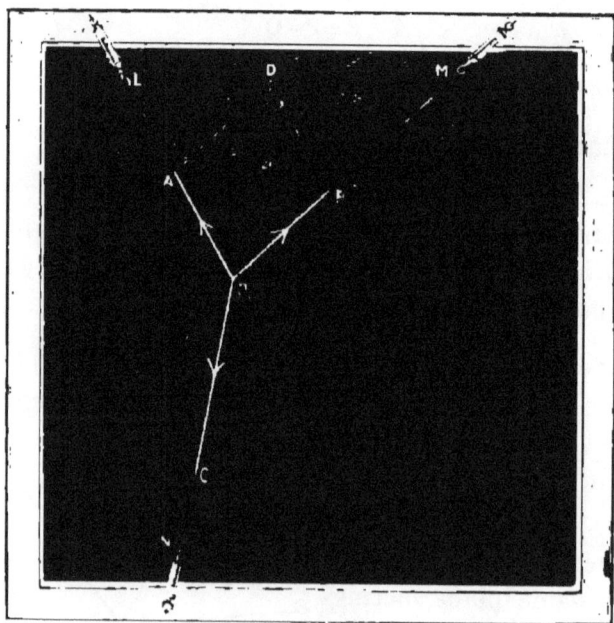

Fig. 13.

Let P denote the force acting along OL,

　　Q denote the force acting along OM,

and　R denote the force acting along ON.

The reading of each balance will give the measure of the force acting along the string attached to it

Take the reading of each balance.

$$P = \quad ?$$
$$Q = \quad ?$$
$$R = \quad ?$$

Draw from the point O a straight line upon the board directly under each string, and, choosing a suitable unit of length to represent a unit of force, lay off on the line under OL, OA containing P units, on the line under OM, OB containing Q units, and on the line under ON, OC containing R units.

By means of a ruler and a pair of compasses construct a parallelogram having OA and OB as adjacent sides.

Draw the diagonal OD, measure its length and determine with a straight edge whether it is in the same straight line with OC.

If the experiment is performed with care it will be found that OD equals OC, and is in one and the same straight line with it.

Therefore, when OA represents the force P and OB the force Q, OD represents a force which is equal and opposite to R; but since P, Q and R are in equilibrium, the resultant of P and Q is equal in magnitude and opposite in direction to R. Therefore OD represents the resultant of P and Q. Hence,

When two forces acting at a point are represented in magnitude and direction by two adjacent sides of a parallelogram, the resultant of these two forces will be represented in magnitude and direction by the diagonal of the parallelogram passing through the point of junction of the two sides which represent the forces.

7. To find the resultant of two forces P and Q which act at right angles to each other.

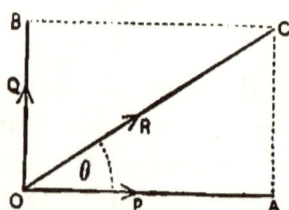

FIG. 14.

Draw the line OA to represent the force P, and the line OB to represent the force Q (Fig. 14).

Complete the parallelogram AOBC.

Then the resultant of P and Q will be represented by OC, the diagonal of the parallelogram.

Let R denote the resultant.

Since the angle OAC is a right angle,

$$OC^2 = OA^2 + AC^2$$
$$= OA^2 + OB^2$$
$$\therefore \quad R^2 = P^2 + Q^2$$
$$\text{or} \quad R = \sqrt{P^2 + Q^2}.$$

If θ denotes the angle which the direction of the resultant makes with the direction of the force represented by OA,

$$\tan \theta = \frac{AC}{OA} = \frac{OB}{OA} = \frac{Q}{P}.$$

EXERCISE VII.

1. Find the greatest and the least resultants of two forces whose magnitudes are 15 grams and 20 grams.

2. Find the greatest and least resultants of two forces whose magnitudes are $P+Q$ and $P-Q$.

3. Find the resultant of forces of 15 pounds and 36 pounds, acting at right angles to each other.

4. Find the resultant of two forces of 12 kilograms and 35 kilograms acting at a point, the one acting north and the other east.

5. The resultant of two forces acting at right angles is 82 pounds. If one of the forces is 80 pounds, what is the other?

6. A force of 5 P acts in a northerly direction, and the resultant of it and another force acting at the same point in an easterly direction is 13 P. What is the other force?

7. The resultant of two forces which are in the ratio 3:4 is 20 pounds when the forces act at right angles to each other. What are the forces?

8. Two forces which are in the ratio 5:12, act at right angles to each other. If the resultant of the forces is 52 grams, find the forces.

9. If two forces acting at right angles to each other are in the ratio $2:\sqrt{5}$, and their resultant is 9 pounds, find the forces.

10. Two forces acting in opposite directions to each other have a resultant of 5 pounds. If they were to act at right angles to each other their resultant would be 25 pounds. Find the forces.

11. Two forces acting in the same direction in the same straight line have a resultant of 34 grams. When these forces act at right angles to each other their resultant is 26 grams. What are the forces?

12. Determine the resultant of the following forces acting concurrently at the same point:—12 pounds N., 24 pounds E., 7 pounds S., and 36 pounds West.

13. A weight is supported by two strings. If the strings make an angle of 90° with each other, and the tension of the one is 9 pounds, while that of the other is 12, what is the weight?

14. A boat is moored in a stream by a rope fastened to each bank. If the ropes make an angle of 90° with each other, and the force of the stream on the boat is 500 pounds, find the tension of one of the ropes if that of the other is 300 pounds.

8. To find the resultant of two forces P and Q inclined to each other at an angle θ.

Draw OA and OB to represent the forces P and Q respectively. Complete the parallelogram AOBC, and join OC. Then the diagonal OC will represent the resultant of P and Q.

Let R denote the resultant of P and Q.

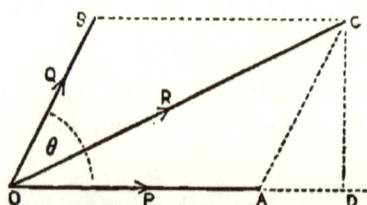
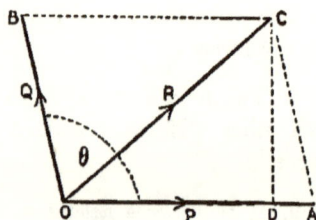

FIG. 15a.　　　　　　　　　FIG. 15b.

Draw CD perpendicular to OA (Fig. 15b), or OA produced (Fig. 15a).

In Fig. 15a.

$$OC^2 = OA^2 + AC^2 + 2\,OA.\,AD \qquad \text{Euc. II., 12.}$$
$$= OA^2 + AC^2 + 2\,OA.\,AC \cos DAC$$
$$= OA^2 + OB^2 + 2\,OA.\,OB \cos \theta$$
since AC = OB
and DAC = θ.

In Fig. 15b,

$$OC^2 = OA^2 + AC^2 - 2\,OA.\,AD \qquad \text{Euc. II., 13.}$$
$$= OA^2 + AC^2 - 2\,OA.\,AC \cos DAC$$
$$= OA^2 + OB^2 + 2\,OA.\,OB \cos \theta$$
since AC = OB
and $\cos DAC = -\cos (180° - DAC)$
$$= -\cos \theta.$$

Therefore,

$$R^2 = \quad P^2 + Q^2 + 2\,PQ \cos \theta.$$

or,

$$R = \sqrt{\{\, P^2 + Q^2 + 2\,PQ \cos \theta \,\}}.$$

EXERCISE VIII.

1. Find the resultant of the following forces :

  (1) 36 pounds and 60 pounds at an angle of 60°.

  (2) 10 pounds and 10 pounds at an angle of 45°.

  (3) 10 pounds and 10 pounds at an angle of 150°.

  (4) 30 pounds and 80 pounds at an angle of 120°.

  (5) 2 pounds and 7 pounds at an angle of 30°.

  (6) 2 pounds and 3 pounds at an angle of 135°.

  (7) 3 pounds and 16 pounds at an angle of 15°.

  (8) 4 pounds and 11 pounds at an angle of 75°.

  (9) P acting toward the west and $P\sqrt{2}$ toward the north-east.

2. Prove that the resultant of two forces, P and $P + Q$, acting at an angle of 120°, is equal to the resultant of two forces, Q and $P + Q$, acting at the same angle.

3. Find the resultant of two forces of 10 pounds and 9 pounds acting at an angle whose tangent is $\frac{4}{3}$.

4. Find the resultant of two forces of 13 pounds and 11 pounds acting at an angle whose tangent is $\frac{12}{5}$.

5. The resultant of two forces which act at an angle of 120° is equal to one of the forces. Find the ratio of the forces.

6. The resultant of two forces which act at an angle of 60° is 13 grams. If one of the forces is 7 grams, find the other.

7. A particle is acted upon by two forces, one of which is inclined at an angle of 80° to the vertical, and the other at an angle of 40° to the vertical and on the other side of it. If one of the forces is 10 pounds, and the combined effect of the two is $2\sqrt{31}$ pounds, find the other force.

8. Two forces which act at an angle of 60° are in the ratio 3:5. If their resultant is 28 pounds, find the forces.

9. Two forces which are in the ratio of 1 to $\sqrt{2}$ act at an angle of 135°. If their resultant is 10 pounds, what are the forces ?

10. Find the magnitude of two forces which have a resultant of $\sqrt{10}$ grams when they act at right angles to each other, and a resultant of $\sqrt{13}$ grams when they act at an angle of 60°.

11. The directions of two forces acting at a point are inclined to each other (1) at an angle of 60°, (2) at an angle of 120°, and the respective resultants are in the ratio $\sqrt{7} : \sqrt{3}$. What is the ratio of the magnitudes of the forces?

12. Two forces of 2 pounds each, acting at an angle of 60°, have the same resultant as two equal forces acting at right angles. What is the magnitude of these forces ?

13. Six posts are placed in the ground so as to form a regular hexagon, and an elastic cord is passed around them and stretched with a force of 50 pounds. Find the magnitude and the direction of the resultant pressure on each post.

14. If one of two forces acting on a particle is 5 kilograms, and the resultant is also 5 kilograms, and at right angles to the known force, find the magnitude and the direction of the other force.

15. The magnitudes of two forces are in the ratio 3:5, and the direction of their resultant is at right angles to that of the smaller force. Compare the magnitudes of the larger force and the resultant.

16. The sum of two forces is 36 pounds, and the resultant, which is at right angles to the smaller force, is 24 pounds. Find the magnitude of each force.

17. Find the resultant of two forces represented by the side of an equilateral triangle and the perpendicular on this side from the opposite angle.

18. The resultant of two forces, P and Q, is $Q\sqrt{3}$, and its direction makes an angle of 30° with the direction of P. Show that P is either equal to Q or 2 Q.

19. Show that when two forces act at a point their resultant is always nearer the greater force, and the greater the angle between the forces the less is their resultant.

20. If a uniform heavy bar is supported in a horizontal position by a string slung over a peg and attached to both ends of the bar, prove that the tension of the string will be diminished if its length is increased.

21. A weight is suspended by means of two strings of equal length attached to points in the same horizontal line. Show that if the lengths of the strings are increased their tension is diminished.

22. Two forces act at a point at right angles to each other, and the magnitude of the smaller is one-half that of the resultant. Show that the angle which this force makes with the resultant is double the angle which the other force makes with it.

23. If the magnitude of one of two forces acting at a point is double that of the other, show that the angle between its direction and that of their resultant is not greater than 30°.

24. If AB and AC represent two forces, and if D is the middle point of BC, show that the resultant of the forces will act along AD and will be represented in magnitude by 2 AD.

25. If D is the middle point of the side BC of a triangle ABC, show that the resultant of the forces represented by the lines AB, AC, DA is represented by the line AD.

26. The side BC of an equilateral triangle ABC is bisected at D, and AD is bisected at O. Two forces, each equal to $\sqrt{7}$ pounds, act along OB, OC. Find the magnitude and the direction of the resultant.

27. ABDC is a parallelogram, and AB is bisected in E. Show that the resultant of the forces represented by AD, AC is double of the resultant of the forces represented by AE, AC.

28. Show that the resultant of the forces represented by AC, DB, the diagonals of the parallelogram ABCD, is represented by 2 AB or 2 DC.

29. If ABC is a triangle and AB is bisected at D, AC at E, and BC at F, show that FA represents the resultant of the forces represented by BE and CD.

30. If two forces acting at a point are represented in magnitude and direction by the sides AB, BC of the triangle ABC, prove that the side AC represents the resultant.

31. Make use of the proposition stated in the last question to solve the following : —

(1) The side BC of an equilateral triangle ABC is bisected at D, and forces are represented in direction and magnitude by BA, BD. Find the magnitude of their resultant, if the force along BD is equal to a weight of 1 pound.

(2) Find the resultant of three forces represented by the sides AB, BC, CD of a rhombus ABCD.

(3) The sides AB and AC of the triangle ABC are bisected at the points D and E. Show that the resultant of the forces represented by DB, BC, CE are equivalent to the resultant of those represented by DA, AE.

(4) If the sides BC, CA, AB, of a triangle ABC are bisected at D, E and F respectively, show that the resultant of the forces represented by AB, AC, BE will be represented by 3 FD.

32. A point is taken within or without a quadrilateral, and lines are drawn from it to the angular points of the quadrilateral. Prove that the resultant of the forces represented by these lines is represented by four times the line joining this point with the middle points of the opposite sides.

33. A straight line is drawn parallel to the base BC of a triangle ABC, cutting AB at the point D, such that AD = 2 BD. If P is any point on the line, prove that the resultant of the forces represented by AP, BP, CP acts in this line.

34. ABCD is a quadrilateral, and AB, BC, CD, DA are bisected at the points E, F, G, H, respectively. Prove (1) that the resultant of the forces represented by AB and DC is represented by 2 HF, 2 that the resultant of the forces represented by EG and HF is represented by AC.

CHAPTER V.

In the preceding chapter the parallelogram of forces was employed to determine the resultant of two forces acting at a point. We shall now apply it to resolving a single force into two components which act in assigned directions.

1. To Find the Components of a Given Force in Two Given
   • Directions.

Let R denote the given force, and a and β the angles which the components make with it.

Draw the line OC (Fig. 16) to represent R; at the point O make the angle COL = a, and the angle COM = β; and from the point C, draw CA parallel to OM, and CB parallel to OL.

Then since AOBC is

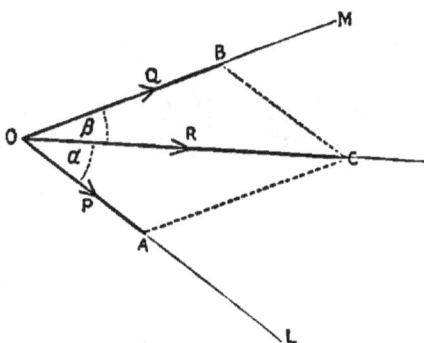

Fig. 16.

a parallelogram and OC represents R, OA and OB will represent the forces P and Q respectively, where P and Q denote the required components.

2. Resolved Part.

When a force is resolved into two components at right angles to each other, each component is called the

Resolved Part of the force in its own direction; that is, the Resolved Part of a force in a given direction is the force in that direction which, together with one at right angles to it, has the given force for a resultant.

The expression **resolved part** calls special attention to one of the components of a force, but it should not be forgotten that another force acts in conjunction with it, and at right angles to it.   For example, if a body has a tendency to acceleration in a north-easterly direction, it has a tendency in a northerly direction accompanied by a tendency in an easterly.   •

3. To Find the Resolved Part of a Given Force in a Given Direction.

Draw the line OC to represent the given force in magnitude and direction (Fig. 17). From O draw the line OL

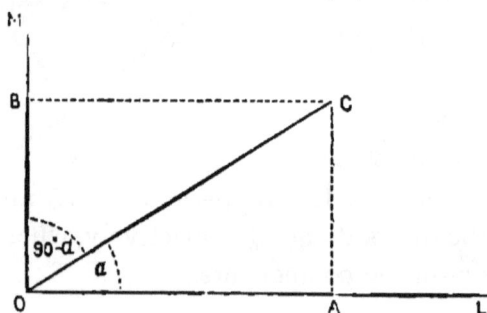

Fig. 17.

in the given direction, and the line OM at right angles to it. From C draw CA parallel to OM, and CB parallel to OL. Then the forces represented by OA and OB at right angles to each other have for their resultant the given force represented by OC.

OA, therefore, represents the resolved part of the given force in the given direction.

Let F denote the given force, and *a* the angle which the given direction makes with the direction of the given force.

Then, $\dfrac{OA}{OC} = \cos AOC = \cos a$

or, $OA = OC \cos a,$

therefore, the resolved part of a force F in a direction making an angle *a* with it

$$= F \cos a$$

Hence,

The resolved part in a given direction is obtained by multiplying the given force by the cosine of the angle between the given force and the given direction.

The component perpendicular to the resolved part

$$= F \cos (90 - a)$$
$$= F \sin a$$

and the two components, therefore, are

$F \cos a,$ and $F \sin a.$

EXERCISE IX.

1. Find the resolved part of a force of 10 pounds in a direction making an angle with the direction of the force of (1) 30°, (2) 45°, (3) 75°.

2. Find the horizontal and the vertical resolved parts of a force of 20 pounds, making an angle of 30° with the horizontal.

3. Find the resolved part S.W. of a force of 12 pounds S.

4. A force of 100 pounds is resolved into two equal forces at right angles to each other. What is the magnitude of either force?

5. The resultant of two forces acting at right angles is 16 pounds, and makes an angle of 30° with one of the components. Find the magnitude of the components.

6. The horizontal resolved part of a force making an angle of 30° with the horizontal is 4 pounds. Find the vertical resolved part.

7. A horse, in towing a canal boat, pulls with a force of 200 pounds. If the tow-rope is horizontal and makes an angle of 5° with the direction of the canal, find the magnitude of the force that would have to be applied in the direction of the canal to draw the boat.

8. A horse draws a load placed upon a sleigh. If he pulls with a force of 100 pounds when the traces make an angle of 10° with the road, with what force must he pull when the traces are parallel with the road? How would a change in the inclination of the road affect the result?

9. Show that the pressure of a perfectly smooth body resting on a perfectly smooth surface is at right angle to the surface.

10. Show that it is possible for a vessel to sail east, against a south-east wind.

11. A force of 12 pounds acts along the side AB of an equilateral triangle. What is the resolved part of this force (1) along the side AC, (2) in a direction parallel to CB?

12. AB represents a force, and a circle is described on AB as diameter. Show that the resolved part of this force in any direction is represented by the chord of the circle drawn in that direction through A.

---

4. To Find the Resultant of any Number of Forces Acting at a Point in given Directions Lying in one Plane.

If the forces act in the same straight line, it is evident that their resultant is the algebraic sum of the forces.

If the forces do not act in the same straight line, let O represent the given point, and through it draw two lines xx' and yy' at right angles to each other.

Let P_1, P_2, P_3 . . . denote the magnitudes of the forces, and a, β, γ . . . the angles which they make with Ox. Draw lines to represent the direction of the forces, as shown in Fig. 18.

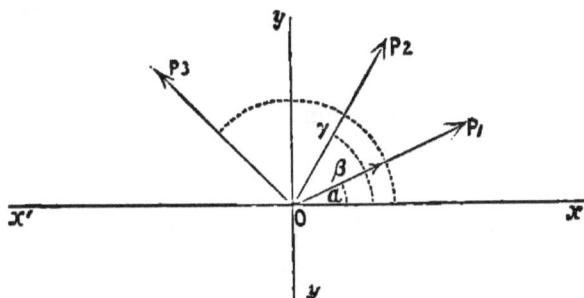

FIG. 18.

Let X denote the algebraic sum of the resolved parts of the given forces in the direction Ox,

Y denote the algebraic sum of the resolved parts of the given forces in the direction Oy

and R denote the resultant of the given forces.

Substituting for each of the given forces its resolved parts in the directions Ox, Oy, we have (Art. 3, page 45),

$$X = P_1 \cos a + P_2 \cos \beta + P_3 \cos \gamma \ . \ . \ . \ . \ . \ . \ . \ .$$
$$Y = P_1 \sin a + P_2 \sin \beta + P_3 \sin \gamma \ . \ . \ . \ . \ . \ . \ .$$

The resultant of the given forces

= the resultant of the equivalent forces substituted for them

= the resultant of X and Y at right angles to each other.

Therefore $R = \sqrt{X^2 + Y^2}$ (Art. 7, page 36), and if θ denotes the angle which the resultant makes with Ox,

$$\tan \theta = \frac{Y}{X}$$

5. Example.

Four forces of 2 pounds, 4 pounds, $6\sqrt{3}$ pounds, and 8 pounds act at a point. If the angle between the first and second is 60°, between the second and third 90°, and between the third and fourth 150°, find the magnitude and the direction of their resultant.

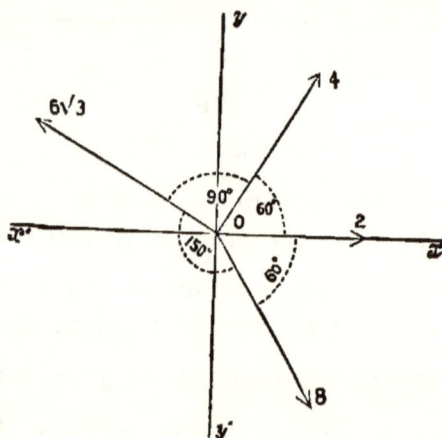

Fig. 19

Let O be the given point, and through it draw two lines xx' and yy' at right angles to each other.

Suppose the 2-pound force to act along Ox, and draw lines to represent the directions of the others, as shown in Fig. 19.

Let X the algebraic sum of the resolved parts of the forces in the direction Ox, and $Y =$ the algebraic sum of the resolved parts of the forces in the direction Oy.

Substituting for each of the forces its resolved parts in the directions Ox, Oy, we have,

$$X = 2 + 4 \cos 60^\circ - 6\sqrt{3} \cos 30^\circ + 8 \cos 60^\circ$$
$$= 2 + 2 - 9 + 4 = -1$$
$$Y = \quad 4 \cos 30^\circ + 6\sqrt{3} \cos 60^\circ - 8 \cos 30^\circ$$
$$= \quad 2\sqrt{3} + 3\sqrt{3} - 4\sqrt{3} = \sqrt{3}$$

The resultant of the four forces

= the resultant of the equivalent forces substituted for them

= the resultant of X and Y at right angles to each other

$$= \sqrt{X^2 + Y^2}$$
$$= \sqrt{(-1)^2 + (\sqrt{3})^2} = \sqrt{4} = 2 \text{ pounds.}$$

If θ denotes the angle which the resultant makes with Ox,

$$\tan \theta = \frac{Y}{X} = \frac{\sqrt{3}}{-1} = -\sqrt{3} = \tan 120^\circ$$
$$\therefore \theta = 120^\circ$$

or the resultant makes an angle of 120° with the first force.

EXERCISE X.

1. Forces of 4 pounds, 2 pounds, and 1 pound act at a point in one plane. Find the resultant when the angle between the first and second, and also between the second and third is 60°.

2. Forces of 4, 8, and $8\sqrt{3}$ pounds act at a point in one plane. If the angle between the first and the second is 60°, and the angle between the first and the third 90°, find the magnitude and the direction of the resultant.

4

3. **Three forces,** 10 pounds, 20 pounds, and 26 pounds, act at a point. If the forces act in the same plane and the angle between the directions of any two of them is 120°, find the magnitude of the resultant.

4. **Four forces,** 20 grams, 20 grams, 10 grams, and 10 grams, act at a point. If the forces act in one plane and the angle between the first and the second is 45°, between the second and the third 75°, and between the third and the fourth 120°, find the resultant.

5. **Forces** of 6 pounds, 9 pounds, and 12 pounds act at a point in directions parallel to the sides of an equilateral triangle taken in order. Find the magnitude and the direction of the resultant.

6. **Forces** P, P, and Q act at a point in direction parallel to AC, CB, AB the sides of an equilateral triangle ABC. Show that their resultant is P + Q.

7. **Five equal forces** of 2 pounds each act along the radii of a circle which are at angular distances 30°, 60°, 90°, 120°, and 150° from a fixed radius. Find the resultant.

8. **At the point O** in the intersection of the diagonals of a square ABCD act forces of 2 pounds along OA, 4 pounds along OB, 3 pounds parallel to CD, and 1 pound parallel to DA. Find their resultant.

9. **Forces** 2 P, $\sqrt{3}$ P, 5 P, $\sqrt{3}$ P, and 2 P act at one of the angular points of a regular hexagon towards the five other angular points. Find the direction and the magnitude of the resultant.

10. **Five equal forces** act on a particle in directions parallel to five consecutive sides of a regular hexagon taken in order. Find the magnitude and the direction of their resultant.

---

6. Conditions of Equilibrium of any Number of Forces in the Same Plane Acting at a Point.

If a number of forces in the same plane act at a point they will be in equilibrium when the resultant is zero.

That is, when

$$R = \sqrt{X^2 + Y^2} = 0 \qquad \text{(Art. 4, page 48).}$$

but the sum of the squares of two real quantities can be zero only when each quantity is separately zero.

Therefore, the forces are in equilibrium when

$$X = 0$$
$$Y = 0.$$

Hence any number of forces acting at a point are in equilibrium when the algebraic sums of the resolved parts of the forces in two directions at right angles separately vanish.

7. Examples.

1. A mass of 3 lbs. is suspended by two strings, one horizontal and the other making an angle of 30° with vertical. Find the tension of each string.

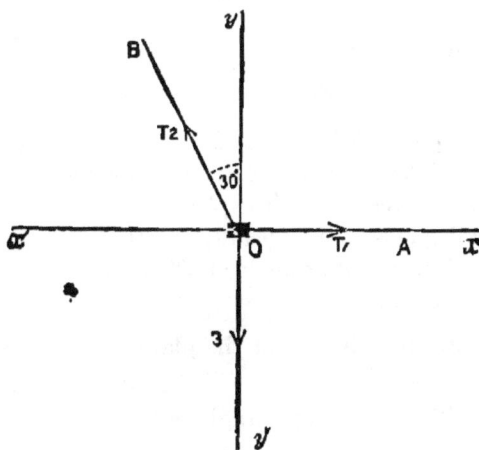

FIG. 20.

Let T_1 and T_2 denote the tensions of the strings, T_1 acting along OA, and T_2 acting along OB. Draw two lines xx' and yy' at right angles to each other in the position shown in Fig. 20.

Three forces keep the mass at rest: its weight, which is 3 pounds, tension T_1, and tension T_2.

Substituting for these forces their resolved parts in the directions Ox and Oy,

$$X = T_1 - T_2 \cos 60^\circ$$
$$= T_1 - \tfrac{1}{2} T_2$$
$$Y = T_2 \cos 30^\circ - 3.$$
$$\tfrac{1}{2}\sqrt{3}\, T_2 - 3$$

But since the forces are in equilibrium

$$X = 0 \text{ and } Y = 0.$$

Hence, $T_1 - \tfrac{1}{2} T_2 = 0$ . . . . (1)

$\tfrac{1}{2}\sqrt{3}\, T_2 - 3 = 0$ . . . . (2)

From (2) $T_2 = \dfrac{6}{\sqrt{3}} = 2\sqrt{3}$ pounds.

Substituting for T_2 in (1)

$T_1 - \sqrt{3} = 0$ or $T_1 = \sqrt{3}$ pounds.

2. A body whose mass is 5 kilograms rests upon a smooth plane inclined at 30° to the horizon, and is acted on by four forces: (1) its weight; (2) the reaction of the plane; (3) a force equal to the weight of 2 kilograms, acting parallel to the plane and upward; and (4) a force P acting at an angle of 30° to the plane. Determine P when the body is at rest.

Let R denote the reaction of the plane. Since the plane is smooth the pressure upon it will be at right angles to it, therefore R will make a right angle with AB.

Represent the directions of the four forces by lines, as shown in Fig. 21.

Through O draw xx' and yy' at right angles to each other.

Substituting for the forces their resolved parts in the directions Ox and Oy,

We have

$$X = 2 + P \cos 30° - 5 \cos 60°$$

$$= 2 + \tfrac{1}{2}\sqrt{3}\,P - \tfrac{5}{2} = \tfrac{1}{2}\sqrt{3}\,P - \tfrac{1}{2}$$

$$Y = P \cos 60° + R - 5 \cos 30°$$

$$= \tfrac{1}{2}P + R - \tfrac{5}{2}\sqrt{3}$$

But since the forces are in equilibrium

$$X = 0 \text{ and } Y = 0$$

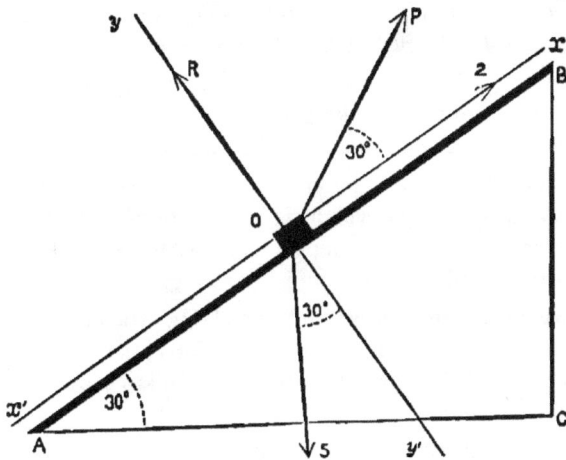

Fig. 21.

Hence,

$$\tfrac{1}{2}\sqrt{3}\,P - \tfrac{1}{2} = 0 \quad . \quad . \quad . \quad . \quad . \quad . \quad (1)$$

$$\tfrac{1}{2}P + R - \tfrac{5}{2}\sqrt{3} = 0 \quad . \quad . \quad . \quad . \quad . \quad (2)$$

From (1)
$$P = \frac{1}{\sqrt{3}} = \frac{\sqrt{3}}{3}$$

Substituting for P in (2)

$$\frac{\sqrt{3}}{6} + R - \frac{5}{2}\sqrt{3} = 0$$

$$R = \tfrac{7}{3}\sqrt{3}$$

Hence P = the weight of a mass of $\dfrac{\sqrt{3}}{3}$ kilograms

and R = the weight of a mass of $\tfrac{7}{3}\sqrt{3}$ kilograms.

EXERCISE XI.

1. Three forces acting at a point are in equilibrium when the angle between the first and the second is 120°, and the angle between the second and the third 150°. If the first force is 20 pounds, what is each of the others ?

2. Three forces acting at a point are in equilibrium. If the angle between any two is 120°, show that the forces are equal.

3. Three forces acting at a point are in equilibrium when the angle between the first and the second is 60°, and the angle between the second and the third is 150°. Compare the forces.

4. Two forces acting on a particle are at right angles, and are balanced by a third force making an angle of 150° with one of them. If the greatest force is 10 pounds, what are the others ?

5. A weight of $10\sqrt{3}$ pounds hangs at the end of a string attached to a peg. If the weight is held aside by a horizontal force, so that the string makes an angle of 30° with the vertical, find the horizontal force and the tension of the string.

6. A weight is hung at the end of a string attached to a peg. If the weight is held aside by a horizontal force, so that the string makes an angle of 60° with the vertical, compare the tension of the string and the weight.

7. A weight of 10 pounds is supported by two strings, one of which makes an angle of 30° with the vertical. If the other string makes an angle of 45° with the vertical, what is the tension of each string ?

8. A string fixed at its extremities to two points in the same horizontal line supports a smooth ring weighing 2 pounds. If the two parts of the string contain an angle of 60°, what is the tension of the string ?

9. A weight of 12 pounds is supported by two strings, each of which is four feet long, the ends being tied to two points in a horizontal line 4 feet apart. What is the tension of each string ?

10. A picture hangs symmetrically by means of a string passing over a nail and attached to two rings fixed to the picture. What is the tension of the string, if the picture weighs 6 pounds and the angle contained by the two parts of the string is 45° ?

11. A uniform bar, the weight of which is 100 pounds, is supported in a horizontal position by a string slung over a peg and attached to both ends of the bar. If the two parts of the string contain an angle of 120°, find the tension of the string.

12. A ball weighing 20 pounds slides along a perfectly smooth rod inclined at an angle of 30° with the vertical. What force applied in the direction of the rod will sustain the ball, and what is the pressure on the rod ?

13. A body, the weight of which is 20 pounds, rests on a smooth plane, inclined to the horizon at an angle of 60°. Find (1) what force acting horizontally will keep the body at rest, (2) the reaction of the plane.

14. A body, the weight of which is 100 pounds, rests on a smooth plane inclined to the horizon at an angle of 30°. What force acting at an angle of 30° to the plane will keep the body at rest ? What is the pressure on the plane ?

15. Two weights of 2 pounds and $\sqrt{6}$ pounds respectively rest, one on each of two inclined planes which are of the same height and are placed back to back. The weights are connected by a string which passes over a smooth pulley at the common apex of the planes. If the first plane makes an angle of 60° with the horizon, find (1) the tension of the string, (2) the pressure on each plane, (3) the inclination of the second plane to the horizon.

CHAPTER VI.

The conditions sufficient for equilibrium when any number of forces in one plane act at a point are given in Art. 6, page 51.

We shall in this chapter consider another statement of these conditions.

I.—Triangle of Forces.

Experiment.

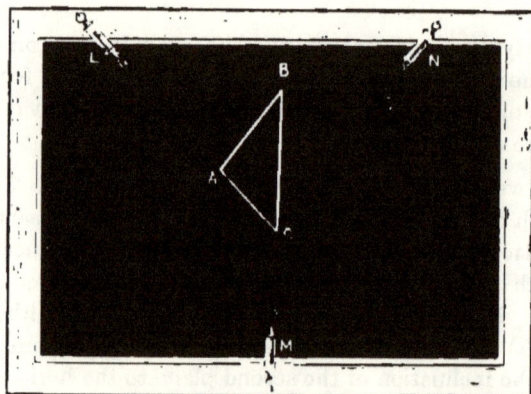

Fig. 22.

Draw on a black-board or a drawing-board any triangle OAB (Fig. 22). Place hooks in the frame of the board at the points L, M, and N ; L being in OA produced, M in BO produced, and N in OC (a line drawn parallel to AB) produced. Hang a spring balance on each hook. Knot together the ends of three strings, and, holding or fastening the

[56]

knot firmly at O, tie each of the free ends of the strings to a spring balance, as shown in the figure, drawing each string so tight that the spring balance to which it is attached will indicate as many units of force as the side of the triangle to which it is parallel contains units of length. Allow the knot to go free. It will be found that the strings remain parallel to the sides of the triangle.

Hence when OA, AB, BO, the sides of a triangle taken in order, represent forces acting at a point O, the forces are in equilibrium.

This proposition is generally known as the **Triangle of Forces.** It may be thus stated.

1. Triangle of Forces.

If three forces acting at a point can be represented in magnitude and direction by the sides of a triangle taken in order, they will be in equilibrium.

This is in reality but another statement of the parallelogram of forces, and the proposition may be derived directly from it.

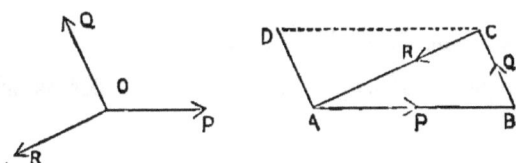

FIG. 23.

Let AB, BC, CA (Fig. 23), the sides of the triangle ABC, taken in order, represent in magnitude and direction the three forces P, Q, R, acting at O.

Complete the parallelogram ABCD.

Since AD and BC are equal and parallel they both represent the same force.

Therefore the forces P and Q will be represented by AB, AD.

But, by the parallelogram of forces, the resultant of the forces represented by AB, AD, is represented by AC.

Therefore the resultant of P, Q, and R, will be represented by AC, CA.

But the forces represented, AC, CA, are in equilibrium.

Hence the three forces, P, Q, and R, are in equilibrium.

The student should carefully observe

(1) That the sides of the triangle are not the lines of action of the forces, but that the forces act at a point.

(2) That the forces must be parallel to the sides of the triangle **taken in order**, P in the direction AB, Q in the direction BC, R in the direction CA, not AC.

(3) That AC represents the resultant of the forces represented by AB, BC; that is, of the forces P and Q.

2. Converse of the Triangle of Forces.

The converse of the Triangle of Forces is also true. It may be thus stated.

If three forces acting at a point are in equilibrium and any triangle is constructed having its sides parallel to the directions of the forces, the forces are proportional to the sides of the triangle taken in order.

This proposition may be verified by an experiment similar to that described in Art. 1, page 56.

Knot the strings together, draw them tight and attach them to the spring balances. Now draw on the board any triangle whose sides are parallel to the directions of the forces, and it will be found that the lengths of its sides are proportional to the magnitudes of forces represented by the spring balances.

II.—Polygon of Forces.

The statement of the conditions of equilibrium of three forces given in the Triangle of Forces may be extended to include any number of forces in the same plane acting at a point.

The proposition is known as the **Polygon of Forces.** It may be thus stated.

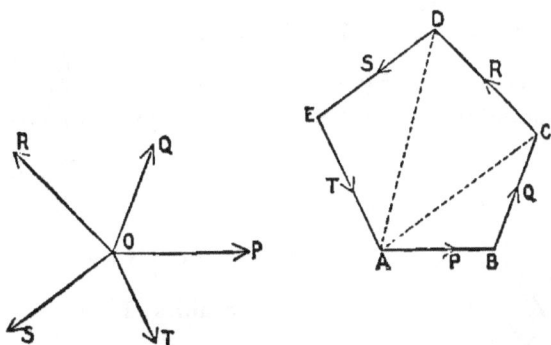

Fig. 24.

If any number of forces acting at a point can be represented in magnitude and direction by the sides of a polygon taken in order, they will be in equilibrium.

Let any number of forces, P, Q, R, S, T, acting at the point O, be represented by AB, BC, CD, DE, EA, the sides, taken in order, of the polygon ABCDE (Fig. 24).

Join AC, AD.

The resultant of the forces represented by AB, BC is represented by AC, Art. 1 (3), page 58.

Similarly the resultant of the forces represented by AC, CD is represented by AD.

And the resultant of the forces represented by AD, DE is represented by AE.

Therefore the resultant of all the forces is represented by AE, EA.

But the forces represented by AE, EA are in equilibrium.

Hence the forces are in equilibrium.

III.—Examples.

1. Three forces, 6P, 7P, 8P, acting in the same plane at a point are in equilibrium. Draw lines which will represent their directions.

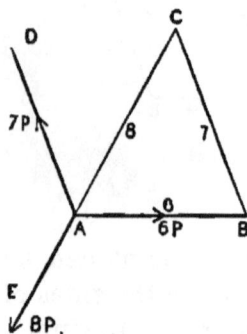

Fia. 25.

Construct a triangle ABC, having the side AB = 6 units of length.

BC = 7 units of length.

CA = 8 units of length. (Fig. 25.)

Then the forces will be represented by the sides of the triangle ABC taken in order.

Suppose the forces to act at the point A.

Draw AD parallel to BC, and produces CA to E.

Then AB, AD, AE will represent the directions of the

forces, because these lines are parallel to the sides of the triangle which represent the forces.

2. A pendulum, consisting of a bob weighing 4 kilograms at the end of a string one metre long, is drawn aside until the bob is 60 cm. from the vertical through the point of support, and is held in position by a horizontal string. Find the forces acting on the bob.

Let A represent the pendulum bob, and B the point to which the string is attached (Fig. 26). Then if T_1 denotes the tension of the pendulum string, and T_2 the tension of the horizontal string, the forces acting on the bob are T_1, T_2, and 4 kilograms, acting in the directions shown in the figure.

Draw the vertical line BC and the horizontal line AC.

Then the forces will be proportional to the sides of the triangle ABC, because the sides of the triangle are parallel to the directions of the forces.

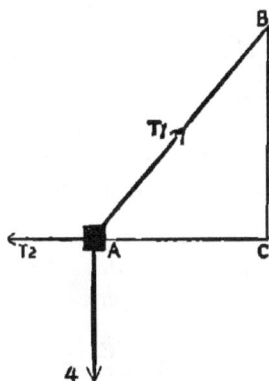

Fig. 26.

Hence

$$\frac{T_1}{4} = \frac{AB}{BC} = \frac{100}{80}$$

or $\qquad$ T = 5 kgm.

and $\qquad$ $$\frac{T_2}{4} = \frac{CA}{BC} = \frac{60}{80}$$

or $\qquad$ $T_2 = 3$ kgm.

3. What horizontal force is necessary to keep a mass of 100 pounds at rest on a smooth inclined plane rising 3 feet in 5? What is the reaction of the plane?

An inclined plane rising 2 feet in 5 is one in which the length AB is 5 feet when the height BC is 3 feet.

Let D represent the body (Fig. 27). If P denotes the horizontal force, and R the reaction of
the plane, the body is kept in equilibrium by

    P, acting horizontally,

    R, acting at right angles to AB,
and 100 pounds, acting vertically
downward. Produce RD to meet
AC in E.

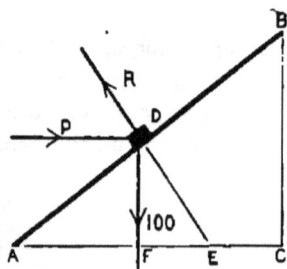

FIG. 27.

Then the forces will be propor-
tional to the sides of the triangle EDF, because the sides of
this triangle are parallel to the directions of the forces.

Hence

$$\frac{P}{100} = \frac{FE}{DF}$$

But $\dfrac{FE}{DF} = \dfrac{BC}{AC}$ Since the triangle EDF is similar
                          to the triangle BAC.

∴ $\dfrac{P}{100} = \dfrac{3}{4}$

or   P = 75 pounds

and $\dfrac{R}{100} = \dfrac{ED}{DF} = \dfrac{AB}{AC} = \dfrac{5}{4}$

or   R = 125 pounds.

EXERCISE XII.

1. Can a particle be kept at rest by each of the following systems
of forces acting at a point ?

    (1) 4 pounds, 3 pounds, 7 pounds.

    (2) 1 gram, 3 grams, 5 grams.

    (3) 4 pounds, 3 pounds, 2 pounds.

    (4) P+Q, P−Q, P, when P is greater than Q.

2. Draw lines to represent the directions of the following forces acting in one place at a point, when each system is in equilibrium :

(1) 4 grams, 5 grams, 3 grams.

(2) Three forces each equal to P.

(3) 2P, P, $\sqrt{3}$P.

(4) 5 grams, 9 grams, 4 grams.

3. Forces 5P, 12P, 13P keep a particle at rest. Show that the directions of two of the forces are at right angles to each other.

4. Find the directions in which three equal forces must act at a point to produce equilibrium.

5. Forces $A + B$, $A - B$, and $\sqrt{2(A^2 + B^2)}$ keep a particle at rest. Show that the directions of two of the forces are at right angles to each other.

6. Forces of 20 pounds, 10 pounds and $10\sqrt{3}$ pounds act on a particle and keep it at rest. Find the angle between the directions of (1) the 20-pound force and the 10-pound force, (2) the 10-pound force and the $10\sqrt{3}$-pound force.

7. The three forces which keep a particle at rest make angles of 60°, 150°, 150° with one another. In what proportions are the magnitudes of the forces ?

8. Two forces acting at a point are at right angles and are balanced by a third force, making an angle of 150° with one of them. If the greatest force is 12 grams, what is the magnitude of each of the others ?

9. A mass of 4 lbs. is suspended from a fixed point by means of a string 35 inches in length, and rests at a distance of 28 inches from the vertical line through the fixed point when acted upon by a horizontal force. Find the horizontal force and the tension of the string.

10. A mass of 10 lbs. is suspended from a fixed point by a string 25 inches in length, and rests 20 inches below a horizontal line drawn through the fixed point when acted on by a horizontal force. Find this force and the tension of the string.

11. A string fixed at its extremities to two points in the same horizontal line supports a ring weighing 20 pounds. If the two parts of the string contain an angle of 60°, find the tension.

12. A mass of 48 lbs. is supported by two strings, which are respectively 6 feet and 8 feet long, and are fastened to two points in the same horizontal line. If the two strings are at right angles to each other, find the tension in each.

13. A mass of 65 lbs. is suspended by two strings, which are respectively 5 feet and 12 feet long, to two points in the same horizontal plane 13 feet apart. Find the tension of each string.

14. A smooth ring sustaining a mass of 48 pounds, slides along a cord fastened at two points lying in the same horizontal line 70 inches apart. If the length of the string is such that the ring rests 12 inches below the horizontal line, find the tension of the string.

15. A picture hangs symmetrically by means of a string passing over a nail and attached to two rings fixed to the picture. What is the tension of the string when the picture weighs 10 pounds, the string is 4 feet long, and the nail 1 foot 6 inches from the horizontal line joining the rings ?

16. A body, the mass of which is 10 lbs., is suspended from a fixed point by a string 25 inches in length. What force acting at right angles to the string will hold the body in a position 20 inches below a horizontal line drawn through the point of suspension ?

17. What force acting parallel to an inclined plane rising 3 feet in 5 feet will support a body the mass of which is 10 lbs.? What is the pressure on the plane ?

18. What force acting horizontally will support a body the mass of which is 12 lbs., on an inclined plane rising 5 feet in 13 feet ? What is the pressure on the plane ?

19. What mass will be supported by a horizontal force of 9 pounds upon an inclined plane rising 3 feet in 5 feet ?

20. A smooth board is fixed at an incline of 1 in 2. A mass of 1 lb. is supported on the board by a string which makes the same angle with the vertical that the board makes with the ground. What is the tension of the string?

21. ABCD is a rhombus. Show that the forces represented by AB, CB, CD and AD are in equilibrium.

22. ABCD is a parallelogram, and three forces acting at a point are represented by AC, BD and 2DA. Show that the forces are in equilibrium.

23. Three forces are represented by AB, AC, two chords of a circle drawn at right angles to each other, and DA, a diameter. Show that the forces are in equilibrium.

24. DC and AB are diameters of a circle. Three forces acting at a point are represented by AB, DC and 2BD. Show that the forces are in equilibrium.

25. Three forces are represented by the lines joining the angular points of a triangle to the middle points of the opposite sides. Show that they are in equilibrium.

26. If AB, AC represent two forces, and D is the middle point of BC, show that the two forces will be balanced by a force represented by 2DA.

27. ABCD is a parallelogram and AB is bisected in E. Show that the forces represented by AD, AC, 2AE, 2CA are in equilibrium.

28. Four forces represented by AB, BC, CD, DE act at a point, and are balanced by a single force represented by AX. What is the position of X?

CHAPTER VII.

1. Friction.

We have, in the problems so far considered, assumed that the surfaces of two bodies in contact were perfectly smooth, and that consequently the stress between the bodies was at right angles to the surfaces in contact. A perfectly smooth surface, like a perfectly straight line or a perfectly fluid body, is but an ideal conception.

Our observations teach us that, while the surfaces of bodies differ widely in smoothness, it is possible to apply to any body resting upon another a small force parallel to the surfaces in contact without producing motion. This shows the existence of a balancing force.

The stress, therefore, between two bodies resting in contact along a plane surface is not necessarily at right angles to this plane, but may be resolved into two rectangular components, one, called the **normal pressure**, acting at right angles to this plane, the other, called **friction**, acting parallel to it.

2. Direction and Magnitude of Friction.

The direction of friction acting upon a body is opposite to that in which motion would take place if there were no friction.

When there is equilibrium the magnitude of friction is equivalent to the magnitude of the least possible force required to maintain equilibrium.

[66]

EXERCISE XIII.

1. A body rests on a horizontal plane and is acted on by a force of 12 pounds making an angle of 60° with the plane. What is the magnitude of the friction called into play?

2. A body rests on a horizontal plane and is acted upon by a force of 20 pounds making an angle of 30' with the plane. What is the magnitude of the friction called into play?

3. A block of wood is in equilibrium on a rough horizontal table when a force of 3 grams acts due north and a force of 4 grams acts due east on it. Find the magnitude of the friction exerted.

4. A body is in equilibrium on a rough horizontal plane when forces of 7 pounds and 8 pounds act upon it. If the forces are parallel to the plane and make an angle of 60° with each other, find the magnitude of the friction exerted.

5. A mass of 60 lbs. rests on a rough plane inclined to the horizon at an angle of (1) 30°, (2) 45°, (3) 60° to the horizon. Find the magnitude of the friction exerted.

6. A mass of 40 lbs. rests on a rough plane inclined to the horizon at an angle of 30°, and a force of 12 pounds acts upon the body parallel to the plane (1) upward, (2) downward. What is the amount of friction called into play?

7. A mass of 100 lbs. rests on a rough plane inclined at an angle of 30° to the horizon and two men push against it, one up the plane with a force of 50 pounds, the other down the plane with a force of 60 pounds. What is the amount of friction exerted?

8. A mass of 100 lbs. rests on a rough plane inclined at an angle of 60° to the horizon, and is acted upon by a horizontal force of 10 pounds (1) toward the plane, (2) away from the plane. Find the magnitude of the friction exerted.

9. A mass of 10 lbs. is in equilibrium on a rough inclined plane of 1 foot in 4 feet. Find the friction exerted when the force of 5 pounds is applied to the mass (1) up the plane, (2) down the plane, (3) horizontally toward the plane.

3. Limiting Friction

Experiment 1.

Arrange apparatus as shown in Fig. 28. The hardwood board AB is about 50 cm. in length and 20 cm. in width, and is provided with some device for holding it at any angle when turned on its hinges (Fig. 30). The upper surface of the board is made as smooth as possible. A vertical scale CD is placed at a given distance AC, say 30 cm., from A.

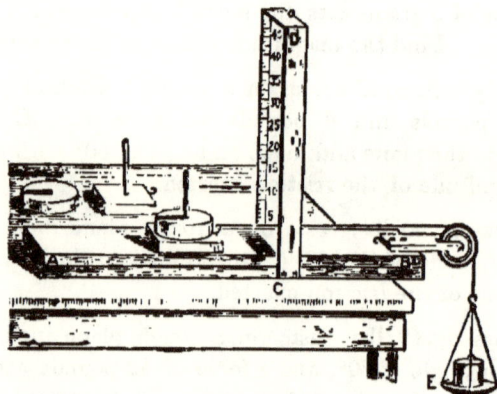

Fig. 28.

Several small pieces of board of different sizes should be provided. These should be cut from a hardwood board, one surface of which has been made as uniformly smooth as possible. Pieces 15 × 5 cm., 15 × 10 cm., 15 × 15 cm. are convenient sizes. Into each a vertical rod for holding weights in position is fastened, and an eye or hook to which a cord can be attached is screwed. It will be found convenient to make them all of the same weight, say one pound, by running lead into holes in their upper surfaces.

Rest the board AB in a horizontal position, and lay on it one of the small pieces of board. Upon this place any weight, say four pounds, and attach the cord and scale-pan as shown in the figure. Place a small weight in the scale-pan and add others in succession until motion takes place. Gently tap the board each time a weight is added to the scale-pan.

It will be found that by adding weights the tension of the string may be increased to a certain limit without producing motion, but if that limit is exceeded the board begins to move.

When a body is just on the point of sliding along another, the body is said to be in limiting equilibrium, and the friction exerted is called limiting friction.

4. Laws of Limiting Friction.

Experiment 2.

Arrange the apparatus used in Experiment 1, Fig. 28. Lay one of the small boards on AB and on it place a weight. Note the total weight supported by the board, including the weight of the board itself. This will then be the measure of the normal stress between the boards. Now add weights to the scale-pan until the board just begins to move. Note the total weight suspended. Repeat the experiment several times, placing different weights on the board.

Denoting the normal pressures by R_1, R_2, R_3, etc., and the limiting frictions by F_1, F_2, F_3, etc., fill up the following table from your observations:

Normal Pressures.	Limiting Frictions.	Quotients.
$R_1 =$	$F_1 =$	$\dfrac{F_1}{R_1} =$
$R_2 =$	$F_2 =$	$\dfrac{F_2}{R_2} =$
$R_3 =$	$F_3 =$	$\dfrac{F_3}{R_3} =$
Etc.	Etc.	Etc.

If the experiment is carefully performed the quotients

$$\frac{F_1}{R_1}, \qquad \frac{F_2}{R_2}, \qquad \frac{F_3}{R_3}$$

will be found to be approximately equal ; that is, the ratio of the limiting friction to the normal pressure between the boards is constant.

Repeat the experiments, using the other pieces of board and making the normal pressures as before, R_1, R_2, R_3, etc. It will be found that, although the shapes and the areas of the surfaces in contact are different, the limiting frictions are approximately F_1, F_2, F_3, etc.

The limiting friction, therefore, is independent of the shape and the area of the surfaces of the boards in contact when the normal pressure remains unaltered.

The experiments of Coulomb and Morin, who have investigated this subject, show that the relations inferred from the above experiments hold approximately for different bodies in contact. The laws must not be looked upon as the statements of absolute truths, but rather as more or less accurate expressions of results determined by careful experiments. They may be thus stated :

Law I.—**The ratio of the limiting friction to the normal pressure is constant when the substances in contact are unaltered.**

Law II.—**The limiting friction is independent of the area and the shape of the surfaces in contact when the normal pressure and the substances in contact remain unaltered.**

To which is sometimes added the following law of kinetic friction, the verification of which does not lie within the province of this work.

Law III.—**When** motion takes place by one body sliding over another, the direction of friction is opposite to the direction of motion; the magnitude of the friction is independent of the velocity, but the ratio of the friction to the normal pressure is slightly less than the constant ratio of the limiting friction to the normal force when the bodies are at rest.

5. Coefficient of Friction.

The constant ratio of the limiting friction to the normal pressure for a particular pair of substances in contact is called the **Coefficient of Friction.** It is generally denoted by μ.

Hence, if F denotes the limiting friction, and R the normal pressure,

$$\frac{F}{R} = \mu$$

or **F** $= \mu$ **R.**

The values of μ differ widely for different substances, depending on the nature of the substances and the degree of polish of their surfaces.

One method of determining the coefficient of friction is given in Experiment 2, page 69. The following is another method of determining this coefficient, and of verifying the laws of limiting friction. If a body is supported upon an inclined plane and the equilibrium is limiting, the forces which keep it at rest are (Fig. 29)

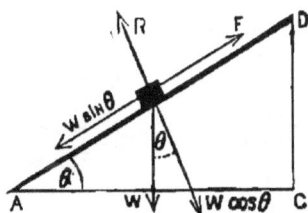

Fig. 29.

Its weight (W), acting vertically downward.

The limiting friction (F), acting along the plane, upward.

The normal pressure (R), acting at right angles to the plane, upward.

If θ denotes the angle DAC,

The resolved part of W along the plane $= W \sin \theta$,

And the resolved part of W at right angles to the plane $= W \cos \theta$. (Art. 3, page 45.)

Substituting for W its resolved parts along the plane and at right angles to it, the forces acting on the body are

(i.) $W \sin \theta$, acting along the plane, downward.

(ii.) F, acting along the plane, upward.

(iii.) $W \cos \theta$, acting at right angles to the plane, downward.

(iv.) R, acting at right angles to the plane, upward.

Since (i.) and (ii.) are at right angles to (iii.) and (iv.), and the body is in equilibrium,

(i.)=(ii.) and (iii.)=(iv.)   Art. 6, page 51.

that is $F = W \sin \theta$

$R = W \cos \theta,$

therefore $\dfrac{F}{R} = \dfrac{W \sin \theta}{W \cos \theta} = \tan \theta = \dfrac{DC}{AC}$

But $\dfrac{DC}{AC}$ may be determined in the following manner with the apparatus used in Experiments 1 and 2.

Experiment 3.

Place one of the small boards on AB and lay a weight upon the board. Gradually raise the end of AB turning it upon the hinges, and at the same time tap it gently. When the small board just begins to slide, fasten AB in position and observe the height CD indicated by the vertical scale.

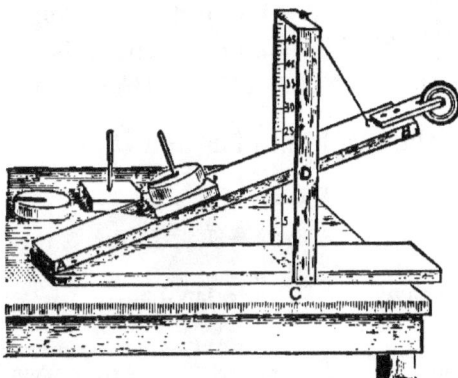

FIG. 30.

Thus $\dfrac{DC}{AC}$ is determined

and since $\dfrac{DC}{AC} = \tan \theta = \dfrac{F}{R}$,

$\dfrac{F}{R}$, or μ, the coefficient of friction of the board is determined

Repeat the experiment, changing (1) the small boards, (2) the weights, (3) both boards and weights.

1. Does the ratio $\dfrac{F}{R}$ remain constant?

2. How does this experiment verify the laws of limiting friction?

6. Limiting Angle of Friction.

The angle which the direction of the resultant of the normal pressure and the limiting friction makes with the direction of the normal pressure is called **the limiting angle of friction, or angle of repose.**

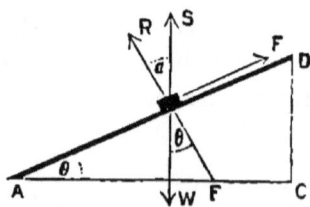

Fig. 31.

For example, if when a body rests upon an inclined plane, R is the normal pressure, F the limiting friction and S the resultant of F and R, the angle a which S makes with R is the angle of friction. (Fig. 31.)

Since S is the resultant of F and R

$$S \sin a = F$$

and $\qquad S \cos a = R.$ Art. 3, page 45.

therefore $\qquad \dfrac{S \sin a}{S \cos a} = \tan a = \dfrac{F}{R}$

But $\qquad \dfrac{F}{R} = \tan \theta.$ Art. 5, page 72.

therefore $\qquad \tan a = \tan \theta$

or $\qquad a = \theta.$

Hence, when a body rests upon an inclined plane under the action of gravity and the reaction of the plane only, and the equilibrium is limiting, **the angle of the inclination of the plane to the horizon is equal to the limiting angle of friction.**

The equality of the angles may be seen directly from the figure.

The body is in equilibrium under the action of the forces F, R and W.

The resultant of F and R is S, which is the reaction of the plane.

The body is, therefore, in equilibrium under the action of W and S.

Therefore W and S are equal and act in opposite directions in the same straight line.

Hence the angle a = the angle θ.

7. Example.

A mass of 12 lbs. rests on a rough plane inclined at an angle of 30° to the horizon. What force must be applied to it at an angle of 30° to the vertical that it may be on the point of moving up the plane, the coefficient of friction being $\frac{1}{2}$?

The forces acting on the body are (Fig. 32)

(i.) Its weight, 12 pounds, acting vertically downward.

(ii.) The normal pressure of the plane, R, acting at right angles to the plane, upward.

(iii.) Friction, $F = \mu R = \frac{1}{2} R$, acting along the plane downward since the body is on the point of moving upward.

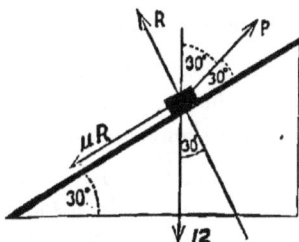

Fig. 32.

(iv.) The required force, P, acting at an angle of 30° to the vertical = 30° to plane.

Resolve the forces along the plane, and at right angles to it.

Then, if X denotes the algebraic sum of the components along the plane, and Y the algebraic sum of those at right angles to it,

$$X = P \cos 30° - \frac{1}{2} R - 12 \cos 60°$$

$$= \frac{1}{2} \sqrt{3} P - \frac{1}{2} R - 6$$

and

$$Y = P \cos 60° + R - 12 \cos 30°$$

$$= \frac{1}{2} P + R - 6 \sqrt{3}$$

Since the body is in equilibrium

$$X=0, \quad Y=0 \qquad \text{Art, 6, page 51.}$$

therefore $\quad \frac{1}{2}\sqrt{3}\,P - \frac{1}{2}R - 6 = 0 \qquad \qquad . \quad (1)$

$$\frac{1}{2}P + R - 6\sqrt{3} = 0 \qquad . \quad . \quad (2)$$

Eliminating R from (1) and (2) by taking (1) × 2+(2)

$$P\left(\frac{1}{2} + \sqrt{3}\right) - 12 - 6\sqrt{3} = 0$$

or $\qquad P = \dfrac{24 + 12\sqrt{3}}{1 + 2\sqrt{3}}.$

EXERCISE XIV.

1. A mass of 10 pounds rests on a rough horizontal plane. If the coefficient of friction is .2, find the least horizontal force which will move the mass. Find also the reaction of the plane.

2. A force of 5 pounds is the greatest horizontal force that can be applied to a mass of 75 pounds resting on a rough horizontal plane without moving it. What is the coefficient of friction?

3. A mass of 10 pounds is resting on a rough horizontal plane, and is acted on by a force which makes an angle of 45° with the plane. If the coefficient of friction is .5, find the force.

4. A body resting on a rough horizontal plane is on the point of moving when acted on by a force equal to its own weight inclined to the plane at an angle of 30°. Find the coefficient of friction.

5. A body placed on a rough plane is just on the point of sliding down when the plane is inclined to the horizon at an angle of (1) 60°, (2) 45°, (3) 30°. What is the coefficient of friction in each case?

6. A body placed on a rough inclined plane is on the point of sliding when the plane rises 3 feet in 6 feet. What is the coefficient of friction?

7. A mass of 20 lbs. rests on a rough plane inclined at an angle of 30° to the horizon. What force must be applied parallel to the plane that it may be on the point of moving up the plane, the coefficient of friction being .1?

8. A body, the mass of which is 30 lbs., rests on a rough inclined plane, the height of the plane being $\frac{2}{3}$ of its length. What force must be applied to the body parallel to the plane that it may be on the point of moving up the plane, the coefficient of friction being .75 ?

9. A rough plane is inclined to the horizon at an angle of 60°. What is the greatest mass which can be sustained upon it by a force of 10 $\sqrt{3}$ pounds acting parallel to the plane, if the coefficient of friction is $\sqrt{3}$?

10. A mass of 14 lbs. when placed on a rough plane inclined to the horizon at an angle of 60° slides down unless a force of at least 7 pounds acts up the plane. What is the coefficient of friction ?

11. A mass of 20 lbs. is on the point of moving up a rough plane inclined to the horizon at an angle of 45° when a horizontal force is applied to it. Find the horizontal force, if the coefficient of friction is .1.

12. A body, the mass of which is 4 lbs., rests in limiting equilibrium when the inclination of the plane to the horizon is 30°. Find the force which acting parallel to the plane will support the body when the inclination of the plane to the horizon is 60°.

13. A body placed on a rough plane inclined to the horizon at an angle of 30° is just on the point of moving upward when acted upon by a horizontal force equal to its own weight. Find the coefficient of friction ?

14. If the smallest force which will move a mass of 3 lbs. along a horizontal plane is $\sqrt{3}$ pounds, find the greatest angle at which the plane may be inclined to the horizon before the mass begins to slide.

15. Show that in order to relieve a horse in drawing a sleigh the traces should be so placed as to make the angle of friction with the ground.

16. How would you place a brick on an inclined plane so that it would be the least likely to *tumble over?* Would it be less likely *to slide down* with one face in contact with the plane than another? Give reasons for your answer.

CHAPTER VIII.

In Chapters xi. and xii. of Part I. we considered the properties of liquids and gases, and described a series of experiments leading up to some of their more important laws. We shall now further consider the laws of fluid pressure and give a number of simple problems in illustration of them.

1. Measure of Fluid Pressure at a Point.

The pressure of a fluid at a point in a plane area is measured when the pressure is uniform over the plane by the force exerted on each unit of area.

For example, if a piston is inserted into the bottom of a vessel filled with water (Fig. 33), and if an upward pressure of 100 pounds is required to hold the piston in position, the pressure of the water on the surface of the piston must equal 100 pounds. Now if the area of the piston is 5 sq. inches, and the square inch is the unit of area, the pressure of the water on each unit-area of the surface of the piston is $\frac{100}{5} = 20$ pounds. The pressure at every point in the surface is said to be 20 pounds per square inch.

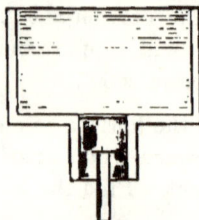

Fig. 33.

If the pressure over the plane is variable, the pressure at a point is measured by the force which

[78]

would be exerted on a unit-area if the pressure were
exerted over the whole unit-area at the same rate as
at the point.

For example, if a piston is inserted into the side of a
vessel filled with water (Fig. 34), the pressure of the
water on the surface of the piston
differs at different points, but is
measured at any point P in the
surface by the force which the
water would exert on a unit-area

FIG. 34.

if the pressure on this unit-area were uniform, and the
same as at P.

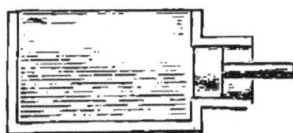

It should be carefully noted that the statement the
pressure at a point is 20 pounds per sq. inch does not
imply that the area of the point is one square inch or
that the pressure upon the point is actually 20 pounds.
The fact is that both the area of the point and the pressure upon it are infinitely small. When it is stated
that the velocity of a train at a certain instant is 20
miles per hour, it is not inferred that the instant is an
hour, or that the distance the train moves in the instant
is 20 miles. It means that at a point of time the train was
moving **at the rate of 20 miles per hour;** that is, if the
train were to continue to move for an hour, and if its
velocity each instant during the hour were the same as
at the given point of time, it would pass over 20 miles.
Similarly, pressure at a point is 20 pounds to the square
inch means that at a certain point in a plane the pressure
is at the rate of 20 pounds to the square inch, that is,
if a surface were one square inch, and the pressure at
every point in it the same as at the given point, the
pressure would be 20 pounds.

2. Pressure at a Point Within the Mass of a Fluid.

To measure the pressure of a fluid at any point within its mass, imagine an indefinitely small rigid plane so placed as to contain the point (Fig. 35). The plane will

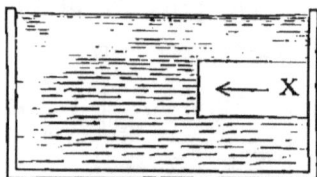

FIG. 35.

in no way affect the pressure of the fluid because it introduces no new forces, nor destroys any of those already existing. Now conceive the fluid removed from one side of the plane; and instead of the pressure of the fluid on that side suppose a force X to keep the plane in position, then the fluid pressure on the other side of the plane must equal X. If the area of surface pressed is a, the pressure at a point in it is $\dfrac{X}{a}$.

3. Laws of Fluid Pressure the Result of the Fundamental Properties of Fluids.

The following laws of pressure which result from the fundamental properties of fluids have been considered in Part I.

Law I.—The pressure at a point of a fluid at rest is perpendicular to any surface with which it is in contact.

From its nature, the tangential resistance to change of shape in any fluid is zero when the fluid is at rest; therefore the strain between a fluid and the surface of a body in contact with it must be perpendicular to that surface.

This may be indirectly illustrated as follows:—

it possible at the point A in the side of a vessel (Fig.

36), let the pressure R of the fluid be not perpendicular. Resolve R into two forces, P and Q, P acting along the side of the vessel, and Q at right angles to it. (Art. 1, page 43.) The effect of Q is balanced by the reaction of the side of the vessel; but since P is unopposed a sliding motion of the particles of the fluid must be taking place in the direction AB.

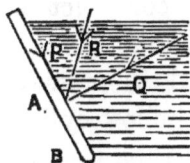

Fig. 36.

This is impossible, because by hypothesis the fluid is at rest; therefore the pressure of the fluid at A does not act in the direction RA. In the same way it can be shown that it does not act in any direction except in one perpendicular to the surface.

Law II.—**Law of transmission of pressure—Pascal's Theory. Pressure exerted anywhere upon a mass of fluid is transmitted undiminished in all directions, and acts with equal intensity upon all equal surfaces, and in directions at right angles to these surfaces.** (Part I., page 104.)

It may be experimentally verified thus:

Take a vessel ABC of any shape (Fig. 37), fill it with any fluid and insert pistons A, B and C; then it will be found that if the piston A is pressed by a force P, to keep them in position, the pistons B and C must be pressed by forces which have the same ratios to P that the areas of the pistons B and C respectively have to the area of the piston A.

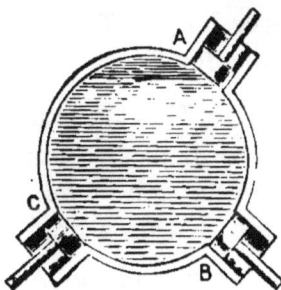

Fig. 37.

6

4. Mechanical Application—Hydrostatic Press.

The equal transmission of fluid pressure is the principle upon which all hydrostatic presses are constructed.

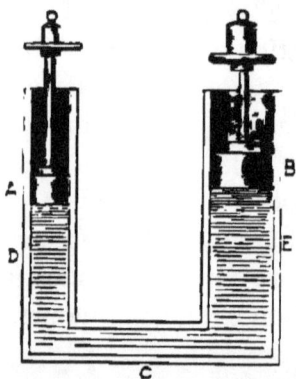

FIG. 38.

Fig. 38 represents one of the simplest forms of these presses. D and E are two hollow cylinders connected by a tube C, and partly filled with water; A and B are two pistons fitted into D and E respectively. Any force applied to A is transmitted through the fluid to B, and the pressures upon A and B are in the ratio of their areas. Thus, if the area of A is one square inch when that of B is ten square inches, a weight of one pound placed upon A will sustain a weight of ten pounds placed upon B.

5. Hydrostatic Paradox.

By decreasing the area of A indefinitely and increasing that of B indefinitely, any force however small applied to A may, by the transmission of pressure through the fluid, be made to support upon B any weight however large. This is sometimes called a "**Hydrostatic Paradox.**"

Law III—Pressure at any point of a fluid at rest is equal in all directions. It follows from the principle of transmission of fluid pressure that the pressure at any point within a fluid mass is the same for all directions.

Let the piston A (Fig. 39) contain some unit of area, for distinctness say one square inch, and let C be any point within the mass of the fluid; imagine it to be the centre of a circular plane, area one square inch and diameter mn. If the piston is pushed inward with a force of P pounds, the pressure on every square inch of the surface of the vessel will be P pounds; and, for the same reason, each face of the circular plane which contains C will be subjected to a force of P pounds. It is evident that

Fig. 39.

on account of the uniform nature of the fluid the magnitude of these forces will remain unchanged if the circular plane is turned round to take different positions in the fluid mass; therefore the pressure at the point which it contains must be the same for all directions.

For experimental verification see Exp. 2, page 109, Part I.

EXERCISE XV.

1. A fluid pressure of 1728 pounds is uniformly distributed over a surface whose area is 3 sq. ft. Find the measure of the pressure at a point in the surface (1) when the unit-area is 1 sq. in., (2) when it is 1 sq. ft., the unit of force in each case being 1 pound.

2. The pressure is uniform over the whole of a sq. yard of a plane area in contact with a fluid, and is 7776 pounds. Find the measure of the pressure at a point (1) when the unit of length is 1 in., (2) when it is 3 in., the unit of force in each case being the pound.

3. The uniform pressure of a fluid over a circular plane, diameter 14 cm., is 770 kgm. ; find the measure of the pressure at a point (1) when the unit-area is 1 sq. mm., (2) when it is 1 sq. dem., if the unit of force is the gram.

4. A rectangular surface, length 50 cm. and width 4 cm., is subjected to a uniformly distributed fluid pressure of 4 kgm. Find the measure of the pressure at a point (1) when the unit of length is 1 mm., (2) when the unit of length is 2 mm., if the unit of force is the gram.

5. If the area of a piston inserted in a closed vessel is $3\frac{1}{4}$ sq. in., and if it is pressed with a force of 35 pounds, find the pressure which it will transmit to a surface of $7\frac{3}{4}$ sq. inches.

6. A closed vessel is filled with fluid and two circular pistons whose diameters are respectively 3 in. and 7 in. inserted ; if the pressure on the larger piston is a pounds, find the pressure on the smaller.

7. The horizontal cross section of the neck of a glass bottle, just capable of sustaining a pressure of 11 pounds to the sq. in., is $2\frac{3}{4}$ sq. in. It is filled with a fluid supposed weightless, and a piston is inserted into the neck. What is the least force that must be applied to the piston to break the bottle ?

8. If the diameter of the small piston (Fig. 38) is 5 cm. and that of the larger one 2.5 metres, and if the small piston is pressed with a force of 8 grams., what force will it transmit to the large piston ?

9. In the same machine the horizontal cross section of the small piston is 3 sq. cm. ; with what force must it be pressed that it may sustain a force of 7.25 kgm. applied to a piston whose horizontal cross section is 7 sq. dcm. ?

10. If the area of the small piston is 2.5 sq. dcm., and if it is pressed with a force of 6.25 centigrams, find the area of the large piston when a pressure of 3.75 grams is transmitted to it.

11. A pressure of 5 tons on the large piston, diameter $2\frac{1}{4}$ feet, transmits a pressure of 2.25 pounds to the small piston. Find the diameter of the small piston.

12. A rectangular box is divided into two compartments by a partition just capable of sustaining a pressure of 1 gram on 1 sq. mm. Both compartments are filled with a fluid supposed weightless, and a piston, area 2 sq. mm., is inserted into the first compartment, and another, area 350 sq. cm., is inserted into the second compartment. If a pressure of 2.5 centigrams is applied to the first piston, what pressure must be applied simultaneously to the second to break the partition?

13. A piston, area $4\frac{1}{2}$ sq. in., is inserted into a rectangular box whose internal dimensions are, length 6 ft., width 3 ft., depth 3 ft. 4 in. If the vessel is filled with water and the piston pressed with a force of 10 oz., find the total pressure on the inside of the box due to the pressure on the piston.

14. A cubical box whose edge is 6 cm. in length is closed by a horizontal lid, and filled with a fluid supposed weightless. An opening of 2 sq. mm. in area is made in the lid, and a piston whose weight is 4 grams is inserted. Find the least weight which must be placed upon the lid to keep it down, if the weight of the lid is 196 grams.

CHAPTER IX.

The forces so far represented as acting on fluids have been pressures applied to their surfaces, and the principles of fluid pressure already investigated are the result of the peculiar constitution of fluids, and are independent of the action of gravity. We shall now establish certain propositions which result from the action of gravity on fluids.

1. The Pressure of a Liquid at Rest Under Gravity is the Same at all Points of the Same Horizontal Plane.

Take any two points A and B (Fig. 40) in the same horizontal plane; consider a very thin cylinder of fluid whose axis is AB.

FIG. 40.

The cylinder is kept at rest by

(i.) The fluid pressures on the curved surfaces, perpendicular to the axis.

(ii.) The weight of the cylinder, acting vertically downward and hence perpendicular to the axis.

(iii.) The fluid pressures on the ends A and B, perpendicular to these ends.

Since (i.) and (ii.) have no tendency to move the cylinder in a horizontal direction, and since the cylinder is at rest, the pressure on the end A must equal the pressure on the end B.

. Now if p denotes the measure of the pressure at a point in the end A, and p_1 the measure of the pressure at a point in the end B, and if a is the area of each end, taken very small that the pressure on each end may be very nearly uniform and of the same intensity as at the middle point, therefore the pressure at the end A is pa and that at the end B is p_1a; but these pressures are equal,

therefore $$pa = p_1a$$
or $$p = p_1$$

2. The Pressure at Any Point of a Liquid at Rest Under Gravity Varies as the Depth. (Part I., page 109.)

Take any point A in the liquid (Fig. 41), and imagine AB drawn vertically to the surface. Consider a thin cylinder of liquid whose axis is AB.

The cylinder is kept at rest by

(i.) The fluid pressures on the curved surfaces of the cylinder, perpendicular to the axis.

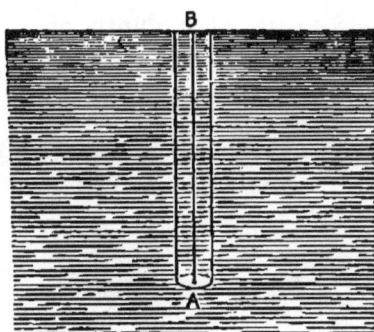

Fig. 41.

(ii.) The weight of the cylinder, acting vertically downward.

(iii.) The fluid pressure on the end A, acting vertically upward.

Since (i.) has no tendency to move the cylinder in a vertical direction, and since the cylinder is at rest, the pressure on the end A must equal the weight of the cylinder.

If p denotes the measure of the pressure at a point in the end A and a is the area of that end, the fluid pressure upon it is pa; and if z is the depth of the point, *i.e.*, the length of the cylinder, and ρ the weight of a unit volume of the liquid, the weight of the cylinder is $\rho a z$; but the fluid pressure on the end A equals the weight of the cylinder,

therefore $$pa = \rho a z$$

or $$p = \rho z$$

Since ρ is constant for the same liquid, any change in z will cause a corresponding change in p. Hence **the pressure at a point varies as the depth.**

For example, if the pressure at a point is 10 pounds per sq. in. at a depth of 2 ft., it will be 20 pounds per sq. in. at a depth of 4 ft., 30 pounds per sq. in. at a depth of 6 ft., etc.

3. The Surface of a Liquid at Rest under Gravity is a Horizontal Plane. (Part I., page 110.)

Take any two points in a horizontal plane within the

Fio 42.

liquid, and imagine vertical lines AC, BD to be drawn to the surface (Fig. 42).

Then the pressure at A $= \rho \times$ AC and the pressure at B $= \rho \times$ BD. (Art. 2);

but the pressure at A = the pressure at B, because the points are in the same horizontal plane (Art. 1),

therefore $\qquad \rho \times AC = \rho \times BD$

or $\qquad\qquad AC = BD$

and since AC is equal to BD and is also practically parallel with it, CD is parallel to AB (Euclid I., 33). Hence C and D must be in the same horizontal plane.

In the case of a large sheet of water on the earth, the surface is not horizontal, but curved. The vertical lines AC and BD when at a great distance apart can no longer be looked upon as practically parallel.

EXERCISE XVI.

1. If the pressure of a liquid at a depth of 14 ft. 3 in. is 6 pounds to the sq. in., find the pressure at a depth of 21 ft. 8 in.

2. If the pressure at a depth of 5.6 metres is 2.8 gm., what is the pressure at a depth of 7.5 cm.?

3. If the pressure on a sq. in. at a depth of 40 cm. is 10 pounds, find the pressure 6 cm. lower down.

4. If the pressure on a sq. mm. at a depth of 5 metres is 2.5 gm., find the pressure 4 dcm. higher up.

5. In two uniform liquids the pressures are the same at depths of 3 and 4 metres respectively. Compare the pressures at depths of 12 and 18 metres respectively.

6. In two uniform liquids the pressures are the same at depths of 1 metre and 1 dcm. respectively. Compare the pressures at depths of 1 dcm. and 1 metre respectively.

7. In three uniform liquids the pressures are the same at depths of 2, 3, and 4 inches respectively. Compare the pressures at depths of 5, 3, and 6 in. respectively.

8. If the pressures at the same depth in two uniform liquids are as 5:6, compare the depths when the pressures are in the ratio 2:3.

9. Find the depth of a pool of water in which a stick 15 feet long stands vertically upon the bottom, if the pressure at the top of the stick is to the pressure at the bottom of it in ratio 3:4.

10. The pressure at the bottom of a stick standing vertically in a pool of water 15 metres deep is to the pressure at the top in the ratio 5:2. Find the length of the stick.

11. Find the measure of the pressure at a point 72 feet below the surface of a pool of water, when the unit of length is 1 in., the unit force 1 pound, and the density of water 62½ pounds per cubic foot.

12. A reservoir of water is 100 metres above the level of the ground floor of a house. Find the pressure of the water at a point in a water-pipe at a height of 10 metres above the ground floor when the unit-area is 1 sq. cm., and the weight of 1 c.cm. of water is 1 gram.

13. The pressure at a point within a body of water under the action of gravity is 100 pounds per square inch. If the weight of a cu. ft. of water is 1000 oz., find the depth of the point below the surface.

14. A reservoir is 200 feet above the level of the ground floor of a house, and the pressure of the water at a point in a faucet in an upper room of the house is 73 }}{ pounds per square inch. Find the height of the faucet above the ground floor if 1 cu. foot of water weighs 1000 oz.

15. Find in grams per square centimetre the pressure at a point due to a column of mercury 1 metre high, the density of mercury being 13.6 grams per cubic centimetre.

16. What must be the height of a column of mercury to exert a pressure of 1 kilogram per square centimetre?

17. The density of sea water is 1.025 grams per cubic centimetre. Calculate the pressure in grams per square centimetre at a depth of 40 metres below the surface of the sea.

18. A spherical boiler 4 feet in height is half full of water and half full of steam. What is the difference, in pounds per square foot, between the pressure at a point at the bottom and at the top of the boiler? (A cubic foot of water weighs 62½ pounds.)

CHAPTER X.

We have seen that when any surface is in contact with a fluid, the fluid exerts a pressure at each point of the surface perpendicular to it. The sum of all these pressures is called the **whole pressure** on the surface immersed.

1. Measure of the Whole Pressure on a Plane.
Experiment 1.

Take a tube A, one end of which is closed by a movable valve

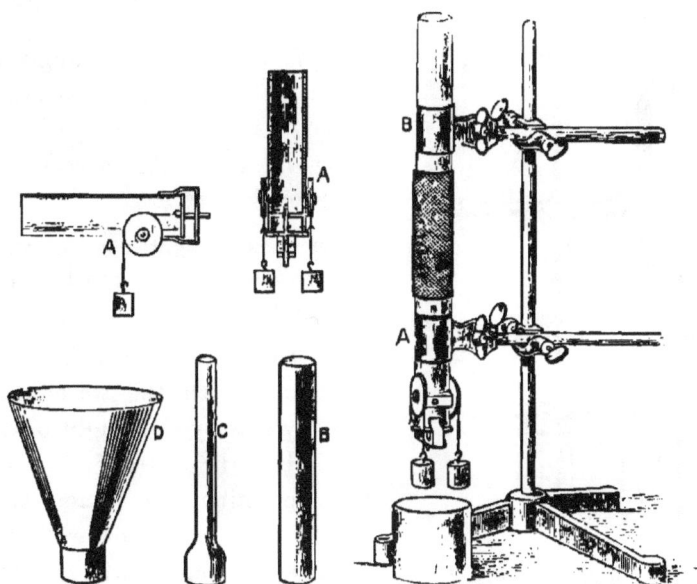

kept in place by two equal weights attached to strings passing over pulleys placed at the side of the tube (Fig. 43).

Place the tube in a vertical position in a holder with the valve down, and, by means of a rubber tube, attach to the upper end of it another glass tube B of the same size. Pour water into the connected tubes until the valve opens and the water falls out. Note the height of the water in the tube when the valve opens. Now remove the upper tube and replace it by others, C and D, of different shapes. Again pour water into the tubes, and observe the height of the water in each case when the valve opens.

How does each of those heights compare with that in the first case ?

Experiment 2.

Repeat Experiment 1, turning the tube A at any angle as shown in Fig. 44, and measuring in each case the height of the water in the tube above the centre of the valve.

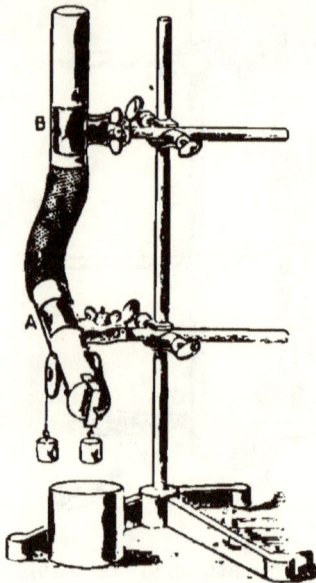

How do the heights compare with those observed in the previous experiment !

Determine the area of the valve in contact with the water, and multiply this by the height of the water above the centre of the valve, thus obtaining the volume of a column of water whose base is the area of the valve, and whose height is the height of the water above the centre of the valve. Find the mass of this water by multiplying the volume obtained by the density of water.

Fig. 44.

the volume obtained by the density of water.

How does the mass of this water compare with the mass of the weights used to keep the valve in position?

If care is taken in the fitting of the apparatus, and if the observations and the calculations are accurately made, it will be found that the valve opens always when the water reaches approximately the same height above the centre of the valve, whatever be the shape of the vessel connected with the tube, or whatever angle the plane of the valve makes with the horizontal.

It will also be found that the mass of a column of water whose base is the area of the valve, and whose height is the height of the water above the centre of the valve, equals the mass of the weights holding the valve in position.

Hence

1. The pressure which any liquid exerts upon a surface in contact with it is independent of the shape of the vessel containing the liquid, or the quantity of the liquid pressing, but depends only on the area of the surface pressed and the vertical distance of its centre of gravity from the surface of the liquid.

2. The whole pressure of any liquid upon a plane in any position in contact with it is equal to the weight of a column of the liquid whose base is the area of the given plane, and whose height is the depth of the centre of gravity of the plane below the surface of the liquid.

The volume of this column of liquid=

Its base × its height,

and its mass=its volume × its density,

but its mass is measured by its weight.

Hence

The whole pressure on a plane = area of the plane × depth of its centre of gravity below the surface of liquid × density of liquid.

2. Examples.

1. Find the whole pressure on a rectangular surface 8 ft. by 6 ft., immersed vertically in water with the longer side parallel to, and 2 ft. below the surface of the water.

Area of the surface pressed $= 8 \times 6 = 48$ sq. ft.

The centre of gravity of the surface is 3 ft. below the upper horizontal side, or 5 ft. below the surface of the water.

Then the volume of the column of water pressing on the surface $= 48 \times 5$ cubic ft.; and, since 1 cubic ft. of water weighs $6\frac{1}{2}$ pounds, the total pressure on the surface $= 48 \times 5 \times 62\frac{1}{2} = 15,000$ pounds.

2. What is the whole pressure exerted against a mill-dam whose length is 100 ft., the part submerged being 10 ft. wide, and the water being 6 ft. deep?

Area of part submerged $= 100 \times 10 = 1000$ sq. ft.

Depth of the centre of gravity of the part submerged from the surface of the water $= \frac{1}{2}$ the depth of the water $= 3$ ft.

Then the volume of the column of water pressing on the part submerged $= 1000 \times 3 = 3,000$ cubic ft. Therefore, the whole pressure $= 3000 \times 62\frac{1}{2} = 187,500$ pounds.

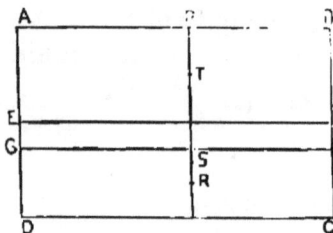

Fig. 45.

3. The flood-gate of a canal is 20 ft. wide and 12 ft. deep, and is placed vertically in the canal, the water being on one side only, and level with the upper edge of the gate; find the whole pressure on (i.) the upper one-half of the gate, (ii.) the lower one-half of the gate, (iii.) the lowest one-third of the gate.

Let ABCD represent the gate (Fig. 45),
     ABFE   "   upper one-half,
     EFCD   "   lower   "
     GHCD   "   lowest one-third,

and let **T**, **S** and **R** denote the centres of gravity of upper one-half, lower one-half and lowest one-third respectively.

(i.) Area of ABFE = 20 × 6 = 120 sq. ft.

Distance PT of the centre of gravity of ABFE below the surface of the water = 3 ft.

Therefore the whole pressure on ABFE = 120 × 3 × 62$\frac{1}{2}$

= 22,500 pounds.

(ii.) Area of EFCD = 20 × 6 = 120 sq. ft.

Distance PS of the centre of gravity of EFCD below the surface of the water = 9 ft.

Therefore the whole pressure on EFCD = 120 × 9 × 62$\frac{1}{2}$

= 67,500 pounds.

(iii.) Area of GHCD = 20 × 4 = 80 sq. ft.

Distance PR of the centre of gravity of GHCD below the surface of the water = 10 ft.

Therefore the whole pressure on GHCD = 80 × 10 × 62$\frac{1}{2}$ = 50,000 pounds.

EXERCISE XVII.

1. Find the whole pressure on a rectangular plane 2 ft. by 4 ft. when immersed in water so that its centre of gravity is 10 ft. below the surface of the water.

2. Find the whole pressure on a rectangular plane 2 metres by 3 metres, immersed horizontally to a depth of 5 metres in water.

3. A rectangular plane 4 ft. by 3 ft. is immersed vertically in water with its shorter sides horizontal, the upper one being 3 ft. below the surface of the water. Find the whole pressure on the surface.

4. A rectangular plane 10 metres by 5 metres is immersed in water with one of its sides horizontal, the upper being 2 metres and

the lower 2.3 metres below the surface of the water. Find the whole pressure on it.

5. A circular surface whose radius is 7 feet, is immersed (1) horizontally, (2) vertically in water. If the depth of the centre of the circle in each case is 8 ft., find the pressure on the surface.

6. The water in a canal lock rises to a height of 10 ft. against one side of a vertical flood-gate whose breadth is 12 ft. Find the pressure against it.

7. Find the whole pressure against a mill dam 40 metres long and 2 metres wide when the water is level with the upper edge of the dam, the lower edge of the dam being 1.8 metres beneath the surface.

8. The water in a canal lock rises to a depth of 20 ft. against a vertical flood gate whose width is 20 ft. Find the pressure on (1) the whole gate, (2) the upper half, (3) the lower half.

9. A rectangular surface 12 ft. by 14 ft. is immersed vertically in water with the longer sides horizontal, the upper being 8 feet below the surface. Find the pressure on (1) the whole surface, (2) the lowest one-quarter, (3) the upper two-thirds, (4) the lowest one-third.

10. A rectangular surface 12 ft. by 8 ft. is immersed in water with its short sides horizontal, the upper being 2 ft. and the lower being 12 ft. below the surface. Find the pressure on (1) the whole surface, (2) the highest one-fifth, (3) the lowest two-fifths.

11. A mill-dam is 8 metres long and 3 metres wide. If the water is level with the top of the dam and the lower edge of the dam is 2 metres below the surface of the water, find the pressure on (1) the whole dam, (2) the upper one-half, (3) the lowest one-quarter.

12. The water in a canal lock rises to a height of 6 metres against one side, and to a height of 4 metres against the other side of a vertical flood-gate whose breadth is 7 metres. Find the whole pressure against the gate.

13. A cube of 1 foot edge is suspended in water with its upper face horizontal and at a depth of $2\frac{1}{2}$ feet below the surface. Find the pressure on each face of the cube.

14. A dyke to shut out the sea is 100 ft. long and is built in courses of masonry 2 ft. high. If the water rises against it to a height of 20 ft., find the pressure on (1) the 2nd course, (2) the 5th course, (3) the 7th course. The courses are to be counted from the bottom.

15. An equilateral triangular plate is immersed vertically in water with an edge 60 cm. in length on the surface. Find the pressure upon one side of it.

16. An isosceles triangular plate whose base is 8 ft. and each of whose equal edge is 5 feet, rests in water in a vertical position with its base horizontal and its vertex in the surface of the water. Find the pressure upon one side of it.

17. A vessel whose base is a square the side of which measures 6 in., contains mercury to the depth of an inch and water to the depth of $10\frac{1}{2}$ in. If the sp. gr. of the mercury is 13.5, find the pressure on the base of the vessel.

18. To what depth must a rectangular surface of 5 sq. ft. in area be sunk in water that it may sustain a pressure of 31,250 pounds?

19. To what depth can a piece of glass whose surface is 60 sq. cm. and just capable of sustaining a pressure of 48 kgm. be sunk in water before it breaks?

20. The cork of a bottle will just sustain a force of 125 pounds to the sq. inch. To what depth can the bottle be sunk before the cork is driven in?

21. Two squares whose sides are 4 and 6 in. respectively are immersed vertically in a liquid, one side of each being horizontal. The first square has its upper side at a depth of 10 in. below the surface. To what depth must the second square be sunk that the pressure on it may be the same as that upon the first?

22. A closed cubical box 12 inches high is filled with mercury of sp. gr. 13.5, and is placed on the flat bottom of a pool of water. What must be the depth of the pool in order that the whole pressure from within on one of the vertical faces may equal the whole pressure from without, assuming that the external and internal faces are of the same area?

7

23. How deep must a 3-inch cube be sunk in water with two of its faces horizontal that the whole pressure on 5 square inches of the bottom may be equal to that on 6 square inches of the top? What then will be the whole pressure on each of the faces?

24. A cube floats with a face level with the surface of a fluid. Find the ratio of the pressures against the bottom and one of the sides.

25. A rectangular box 2 cm. long, 1.5 cm. wide, and 8 mm. deep is filled with water. Find the pressure on (1) the bottom, (2) a side, (3) an end.

26. Compare the pressures on the bottom and a side of a cubical vessel filled with any fluid. Why is the result the same as that of question 24?

27. A cubical vessel is filled with two liquids whose specific gravities are 1 and 0.8 respectively. They do not mix, and their volumes are equal. Find the ratio of the pressure on the upper to that on the lower half of one of the vertical faces of the cube.

28. A cylindrical vessel, height 200 cm. and radius of base 70 cm., is filled with water. Find the pressure on (1) the bottom, (2) the curved surface.

29. A cylindrical vessel with smooth internal surfaces 40 cm. high and 14 cm. in diameter, is filled with water and closed by a piston weighing 5 kgm. Find the pressure on (1) the bottom, (2) the curved surface.

30. A vessel is in the shape of a pyramid which is 4 ft. high and has a square base, each edge of which is 6 feet. Find the pressure on (1) the base, (2) a side, when it is filled with water. If the vessel is supposed weightless, find the pressure on the table upon which the vessel stands.

31. A conical vessel, supposed weightless, 8 cm. high and radius of base 6 cm., is filled with water. Find the pressure on (1) the base, (2) the curved surface, (3) the table upon which the vessel stands.

32. A piston, 2 sq. mm. in area and weighing 10 grams, is inserted into the upper side of a closed cubical box, each edge of

which measures 8 cm. If the box is filled with water, find the whole pressure on the entire internal surface of the box.

33. A cylindrical vessel 1.2 metres high and 1.4 metres in diameter is filled with water. Find the pressure on (1) the base of cylinder, (2) the upper one-half of curved surface, (3) the lowest one-third of curved surface.

34. A conical vessel with its vertex dowrward, is filled with water. Find the total pressure on its curved surface, the diameter of the base being 1 metre and the height being 1.2 metres.

35. To what depth must a liquid be poured into a rectangular box, the base of which is a square whose side is 4, that the sum of the pressures on the sides may be four times the pressure on the base?

36. Find the height of a cylinder, the diameter of which is 12 in. that the pressure on the curved surface may be three times the pressure on the base, when the cylinder is filled with any liquid.

37. A tube whose internal cross-section is 1 sq. cm. opens freely into a water tank whose internal cross-section is 4 sq. m. What pressure must be exerted against a piston which works in the tube by the water rising in the tank to a height of 4 metres above the level of the piston?

38. A rectangular vessel 80 cm. long, 20 cm. wide, and 60 cm. deep, supposed weightless, is placed on a horizontal table. Into its upper face is let perpendicularly a straight tube which rises to a height of 2 metres above this face, the internal cross-section of the tube being 1 sq. cm. The vessel and the tube are filled with water. Find the pressure on (1) the bottom of vessel, (2) a side, (3) an end, (4) the upper surface, (5) the table.

39. A cylindrical vessel, radius of base 8 in. and height 12 in., is placed vertically on a horizontal table. Into its upper end is let perpendicularly a tube 20 in. in length, the horizontal cross-section of which is 1 sq. in. Compare the pressure on the base and on the curved surface, when the cylinder and the tube are filled with any liquid.

40. A cubical vessel supposed weightless, whose edge is 4 in. in length, is placed on a horizontal table. Into its upper surface is

let perpendicularly a straight tube, the horizontal cross-section of which is 1 sq. in. Vessel and tube are filled with a liquid. Find the length of the tube (1) that the pressure on the side of the cube may be four times the pressure on the base before the tube was inserted, (2) that the pressure on the table may be twice the pressure on the base before the tube was inserted.

41. A rectangle, length 6 in., breadth 4, remains immersed vertically in a liquid with one of its long sides in the surface of the liquid, divide it by a horizontal line into two parts upon which the pressures are equal.

---

3. Equilibrium of Liquids of Unequal Density in a Bent Tube.

If two liquids of unequal density which do not mix are poured into a bent tube, they rest in equilibrium in the position shown in Fig. 46, where A and B represent their free surfaces and C their common surface. Let ρ_1 and ρ_2 be the densities of the liquids, and ac and bc the heights of their free surfaces above the horizontal plane CD drawn through C their common surface.

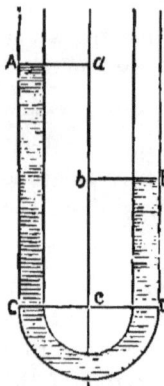

Fig. 46.

Since the liquids are in equilibrium

The pressure at C = the pressure at D.

But the pressure at $C = \rho_1 \times ac$

and the pressure at $D = \rho_2 \times bc$

$$\therefore \quad \rho_1 \times ac = \rho_2 \times bc$$

$$\text{or} \quad \frac{\rho_1}{\rho_2} = \frac{bc}{ac},$$

Hence

When the liquids are in equilibrium, their densities are inversely as the heights of their respective columns above their common surface.

EXERCISE XVIII.

1. Two liquids which do not mix are contained in a bent tube. If their specific gravities are 1.2 and 1.8 respectively, and the height of the first above their common surface is 15 inches, find the height of the other.

2. In a bent tube a column of mercury (sp. gr. 13.6) is balanced by a column of alcohol (sp. gr. .8) if the height of the former is (1) 4 cm., (2) 10 cm., (3) 15 cm., what in each case is the height of the latter ?

3. Two tanks are connected by a pipe. Into one tank is poured salt water (sp. gr. 1.03), and into the other petroleum oil (sp. gr. .5). The oil is found to be 5 ft. above their common surface. Find the height of the water.

4. Mercury and ether are poured into a bent tube. The mercury stands 5.25 cm. when ether stands 100 cm. above their common surface. If the density of the ether is .715 grams per cubic centimetre, what is the density of the mercury?

5. Two liquids that do not mix are contained in a bent tube. The difference of their levels is 40 cm. and the height of the denser above their common surface is 70 cm. Compare their densities.

6. If water and a denser liquid which does not mix with it is placed in a U-tube, the internal cross-section of which is 1 sq.cm., the difference of their levels is found to be 5 cm., and the height of the liquid above their common surface is 10 cm. What is the height of the water ?

7. The lower ends of two vertical tubes whose cross-sections are 1 and .1 inches respectively, are connected by a tube. The tubes contain mercury (sp. gr. 13.6). What volume of water must be poured into the larger tube to raise the level of the mercury in the smaller tube by one inch ?

8. A U-tube has mercury (sp. gr. 13.6) poured into it until the surface of the mercury is half-way up each tube. Water is poured into one branch until it is filled. How high will the mercury rise in the other branch ?

9. A U-tube, having its two arms equal, has mercury poured into it until each surface is 6 inches below the top of the tube. Water is poured into one branch and alcohol (sp. gr. .8) into the other to fill the tube. Find the length of the tube occupied by each liquid.

10. Water is poured into a U-tube, the branches of which are 6 inches long, until they are half full. As much oil (sp. gr. ⅔) is poured into one of the branches. What length of the tube does it occupy?

11. A uniform bent tube consists of two vertical branches and of a horizontal portion uniting the lower ends of the vertical portions. Enough water is poured in to occupy 6 cm. of the tube and then enough oil (sp. gr. ⅘) to occupy 5 cm. is poured in at the other end. If the length of the horizontal part of the tube is 2 cm., find where the common surface is situated.

CHAPTER XI.

If a body is immersed in a liquid, every point of its surface will be subjected to a pressure which is perpendicular to the surface at that point, and which varies as the depth of that point below the surface of the liquid. When these pressures are resolved into horizontal and vertical components, the horizontal components equilibriate each other, and since the pressure on the lower part of the body is greater than that on the upper part, the resultant of all the forces acting upon the body must be vertical and act upward. This force is termed the **resultant vertical pressure** or **buoyancy of the fluid.**

1. Measure of the Resultant Vertical Pressure.

In Experiment 2, page 111, Part I., it was shown experimentally that the buoyant force exerted by a fluid on a body immersed in it is **equal to the weight of the fluid equal in volume to the body,**

or,

A body when weighed in a fluid loses in apparent weight an amount equal to the weight of the fluid which it displaces.

This conclusion may be arrived at in another way.

Let A be a body which is either wholly immersed

[103]

(Fig. 47b) or which has the part abcd (Fig. 47a) immersed in a fluid, and suppose the body removed, and its place

Fig. 47a.

abcd filled up with fluid. Since the fluid body in the space abcd is of the same density as the surrounding fluid, it will remain in equilibrium. The forces acting upon it are:

Fig. 47b.

(i.) Its weight.

(ii.) The resultant fluid pressure upon its surface.

Therefore the resultant fluid pressure equals the weight of the fluid abcd. Since the weight of the fluid acts vertically downward through its centre of gravity, the resultant pressure must act vertically upward through the same point. If now we suppose the solid body A to replace the fluid in the space abcd, it is evident that the resultant pressure of the fluid will thrust it upward with the same force, that is, with a force equal to the weight of the fluid displaced by the solid.

EXERCISE XIX.

1. A cu.ft. of marble which weighs 300 pounds is immersed in water. Find (1) the resultant fluid pressure upon it, (2) its weight in water.

2. A cu.in. of a metal weighs 5 pounds in air. Find its weight in water.

3. Three and one-half c.dcm. of a substance weigh 6 kgm. Find the weight when immersed in water.

4. Six and three-fourths c.cm. of a substance weigh 18.5 grams. Find the weight when immersed in a liquid twice as heavy as water.

5. A body whose volume is $2\frac{2}{3}$ cu.ft. weighs 420 pounds. Find its weight when $\frac{4}{5}$ of its volume is immersed in water.

6. A substance whose volume is $3\frac{1}{2}$ c.dcm. weighs $7\frac{3}{4}$ kgm. Find its weight when $\frac{2}{3}$ of its volume is immersed in a liquid one-half as heavy as water.

7. Five c.metres of a metal weigh *in vacuo* 20,000 kgm. With what force would it be buoyed up if it were suspended in air? Find its weight in air. 1 c.cm. of air weighs .0013 grams.

8. Three and one-half cu.ft. of granite weigh 750 pounds in water. Find its weight.

9. A body whose volume is $4\frac{1}{2}$ c.dcm. weighs in water $5\frac{1}{2}$ kgm. Find its weight.

10. A substance whose volume is m cu. feet weighs n pounds in air. Find its real weight, that is its weight *in vacuo*, if a cu. inch of air weighs p oz.

11. Seven and one-half c. metres of a substance weigh 9,000 kgm. when $\frac{2}{3}$ of its volume is immersed in water. Find its weight.

12. A body, the volume of which is $5\frac{2}{3}$ cu. feet, weighs 75 pounds when $\frac{3}{17}$ of its volume is immersed in a liquid $\frac{2}{3}$ as heavy as water. Find its weight.

13. Find the volume of a body that weighs 10 kgm. in air and 8 kgm. in water.

14. Find the volume of a cube of iron which weighs 2,800 pounds in air and 2,425 pounds in water.

15. Find the edge of a cube of lead which weighs 90.8 kgm. in air and 84.8 kgm. when $\frac{3}{4}$ of its volume is immersed in water.

16. A cu. in. of one of two liquids weighs $\frac{3}{4}$ oz., and of the other $\frac{1}{3}$ oz. A body immersed in the first weighs 7 oz., and in the second 12 oz. Find the weight and the volume of the body.

17. A c.cm. of water weighs 1 gram and a c.cm. of air weighs 0.0013 grams. A body weighs 100 grams in air and 40.078 grams in water. Find (1) its weight *in vacuo*, (2) its volume.

18. The mass of a piece of limestone (sp. gr. = 2.637) is 256.34 gm. What is its apparent weight in water ?

19. The apparent weight of a mineral when weighed in water is 195.46 gm. If its specific gravity is 2.678, what is its mass ?

20. Find the apparent weight of 5 c.cm. of gold (sp. gr. = 19.3) in mercury (sp. gr. = 13.6).

21. A vessel of water is placed in one scale-pan of a balance and counterpoised. Will the equilibrium be disturbed if a person dips his finger into the water without touching the sides of the vessel ? Explain fully.

22. A vessel containing water is placed in one scale-pan of a balance, and balanced by weights in the other scale-pan. A piece of lead (sp. gr. = 11.4) weighing 17.1 gm. is suspended by a string from a fixed support, and is placed totally immersed in the water. What additional weight must be used to restore the equilibrium ? In which scale-pan must it be placed ?

23. Water floats upon mercury, whose specific gravity is 13, and a mass of platinum, whose specific gravity is 21, is held suspended by a string so that $\frac{19}{21}$ of its volume is immersed in the mercury and the remainder of its volume in the water. Prove that the tension of the string is half the weight of the platinum.

24. Eight cubic centimetres of a metal whose specific gravity is 6, and a certain volume of platinum whose specific gravity is 21 are connected by a fine thread passing over a smooth pulley, and rest

in equilibrium when both bodies are immersed in water. By how much must the metal be diminished in volume to preserve equilibrium (1) when the metal is removed from the water, (2) when both the metal and the platinum are removed from the water?

___

2. Conditions of Equilibrium of a Body acted upon by Fluid Pressure.

When a body is placed in a fluid and left to itself two forces act upon it.

(i.) The force of gravity, acting vertically downward through the centre of gravity of the body.

(ii.) The resultant fluid pressure, acting vertically upward through the centre of buoyancy (the centre of gravity of the fluid displaced) = the weight of the fluid displaced.

If the body is in equilibrium under the action of these forces, they must be equal and act in opposite directions in the same straight line. The following are, therefore, the conditions of equilibrium when a body floats wholly or partially immersed in a fluid.

1. **The weight of the body equals the weight of the fluid displaced by it.**

2. **The centre of gravity of the body is in the same vertical line as the centre of buoyancy.**

___

EXERCISE XX.

1. A cubic in. of pine floats with $\frac{5}{7}$ of its volume in water. Find its weight.

2. A c.cm. of poplar floats with $\frac{m}{n}$ of its volume out of water. Find its weight.

3. The weight of $2\frac{1}{4}$ cu. feet of elm is 124 pounds. What part of its volume will be immersed if it is allowed to float in water?

4. The weight of $6\frac{3}{4}$ c.dem. of cork is $1\frac{2}{3}$ kgm. If it is allowed to float in water, how many cu.dem. will remain above the surface?

5. A piece of wood weighing 100 pounds floats in water with $\frac{2}{3}$ of its volume above the surface. Find its volume.

6. If a piece of ash (sp. gr. $= .8$) is allowed to float in water, what part of its volume will be immersed?

7. A cylinder 12 in. long made of larch (sp. gr. $= .53$) floats in water. How many inches will remain out of water?

8. If a body whose specific gravity is 4 float in a liquid whose specific gravity is 5, what portion of the body will be immersed?

9. Seven and one-half cu.ft. of poplar floats with $\frac{2}{3}$ of its volume out of a liquid (sp. gr. $= .8$). Find its weight.

10. A piece of pine weighs $a\,n$ grams and floats with $\frac{a}{n}$ of its volume in water. Find its volume.

11. What is the least force which must be applied to a cu.ft. of larch which weighs 30 pounds that it may be wholly immersed in water?

12. A c.dem. of cork, weighing 480 grams, floats just immersed in water, when prevented from rising by a string attached to the bottom of the vessel containing the water. Find the tension of the string.

13. A cylindrical cup weighs 35 grams, its external radius being $1\frac{3}{4}$ cm., and its height 8 cm. If it be allowed to float in water with its axis vertical, what additional weight must be placed in it that it may sink?

14. A cylinder of wood, 8 in. long and weighing 15 pounds, floats vertically in water with 3 in. of its length above the surface. What is the tension of the string which will hold it just immersed in water?

15. A cylindrical buoy, 10 ft. long and horizontal cross-section 4 sq.ft., weighs 800 pounds. If it be anchored with one-half of its

volume in sea-water. Find the tension of the anchor chain (1 cu.ft. of sea-water weighs 64½ pounds).

√ 16. A cube floats in water with its upper face in the surface when a weight of 62½ pounds is placed upon it. If the cube rises 1 in. when the weight is removed, find the edge of the cube.

17. A cylinder floats vertically in water, and sinks 2 in. when a weight of 250 oz. is placed upon it. Find the horizontal cross-section of the cylinder?

18. A body floats with one-third of its volume immersed in a liquid. If it requires a weight of 10 pounds to cause it to float with one-half of its volume immersed, find its mass.

19. The area of the cross-section of a ship at the water-line is 10,000 sq. ft. What additional load will sink it 2⅔ in.?

— 20. A cylinder floats vertically in a liquid. Compare the forces necessary to raise it 3 in. and to depress it 3 in.

21. What is the least weight that must be placed upon a cu. ft. of cork (sp. gr. =.25) that it may float totally immersed in a liquid whose specific gravity is .9?

22. What is the least weight that must be placed upon a piece of wood weighing 20 pounds and floating with ¾ of its volume immersed in a liquid whose specific gravity is 1.5 that it may be totally immersed?

23. A cylinder of cork weighs 10 grams, and its specific gravity is .25. Find the least force that will immerse it (1) in water, (2) in a liquid whose specific gravity is .75.

24. A body (sp. gr. =.5) floats on water. If the weight of the body is 1 kgm., find the number of cubic centimetres of it above the surface of the water.

25. If a cube float on water with one of its faces horizontal, and a body the mass of which is 9 grams when placed upon it makes it sink 1 cm., find the size of the cube.

26. A cylinder of wood floats in water with three-fourths of its volume above the surface; when a cylinder of metal half as large again is attached to the first, the two float just immersed. Compare the densities of the wood and the metal.

27. A hollow cubical metal box of which the length of an edge is one inch and the thickness one-eighteenth of an inch will just float in water when a piece of cork of which the volume is 4.34 c. in. and the specific gravity .5, is attached to the bottom of it. Find the density of the metal.

28. A cylinder of larch 19 cm. in height is joined to a cylinder of iron (sp. gr. = 7.8) 1 cm. in height so as to form one cylinder 20 cm. in height. This is found to float in water with 2 cm. projecting above the surface. Find the density of larch.

29. A rod of uniform section is formed partly of platinum (sp. gr. = 21) and partly of iron (sp. gr. = 7.5), the platinum portion being 2 in. long. What will be the length of the iron portion when the whole floats in mercury (sp. gr. = 13.5) with the iron 1 in. above the surface?

30. A liquid (sp. gr. = 1.6) is poured into a vessel containing mercury (sp. gr. = 13.1), and a cylinder of zinc (sp. gr. = 7) allowed to sink through the liquid floats with its axis vertical in the mercury. If the cylinder is 5 dcm. long, find the length of the part immersed in the mercury.

31. Taking 7.67 pounds as the weight of 100 cu. ft. of air, find approximately the volume of hydrogen (sp. gr. compared with air ·07) which a balloon must contain in order that its lifting power may be equal to a weight of 713 lbs.

32. A body floats in a fluid (sp. gr. = .9) with as much of its volume out of the fluid as would be immersed if it floated in a fluid (sp. gr. 1.2). Find the specific gravity of the body.

33. A cubical block of wood (sp. gr. = .6) whose edge is 1 foot floats, with two faces horizontal, down a fresh water river out to sea, where a fall of snow takes place, causing the block to sink to the same depth as in the river. If the specific gravity of the sea water is 1.025, find the weight of the snow on the block.

34. A ship, of mass 1000 tons, goes from fresh water to salt water. If the area of the section of the ship at the water line is 15000 sq. ft., and her sides vertical where they cut the water, find how much she will rise, taking the specific gravity of sea water as 1.026.

35. A piece of iron (sp. gr. $= 7.5$), the mass of which is 26 lbs., is placed on the top of a cubical block of wood floating in water, and sinks it so that the upper surface of the wood is level with the surface of the water. The iron is then removed. Find the mass of the iron that must be attached to the lower surface of the wood so that the top may be as before in the surface of the water.

36. A body floats in water contained in a vessel placed under an exhausted receiver with one-half its volume immersed. Air is then forced into the receiver until its density is 80 times that of the air at the atmospheric pressure. Prove that the volume immersed in the water will then be $\frac{4}{5}$ of the whole volume, assuming the specific gravity of air at the atmospheric pressure to be .00125.

37. An accurate balance is completely immersed in a vessel of water. In one scale-pan some glass (sp. gr. $= 2.5$) is being weighed and is balanced by a one-pound weight whose sp. gr. is 8, which is placed in the other scale-pan. Find the real weight of the glass.

38. A cubical box, the volume of which is 1 cu. ft., is three-fourths filled with water, and a leaden ball, the volume of which is 72 cu. in., is lowered into the water by a string. Find the increase in pressure on (1) the base, (2) a side of the box.

39. A cylindrical bucket, 10 in. in diameter and one ft. high, is half filled with water. A half hundredweight of iron is suspended by a thin string and held so as to be completely immersed in water without touching, the bottom of the bucket. Subsequently the string is removed and the iron allowed to rest on the bottom of the bucket. By how much will the pressure on the bottom be increased in each case by the presence of the iron? (A cu. ft. of iron weighs 440 lbs.)

CHAPTER XII.

INSTRUMENTS AND MACHINES.

I.—The Barometer.

The method of measuring the pressure of the atmosphere upon any surface by measuring the weight of a counterbalancing mercury column was considered in Exp. 9, page 121, Part I. This is the principle of action of all mercury barometers. The following is a description of one of the most common forms of the instrument.

1. Cistern Barometer.

A glass tube A (Fig. 48) about one centimetre in diameter and 800 mm. long, one end being closed and the other narrowed to a small opening, is filled with mercury and placed in a vertical position with the open end beneath the surface of the mercury in a small glass cistern B. This cistern has usually a flexible leather bottom which can be moved up or down by a screw C. A scale is attached to the side of the tube by means of which the height of the surface of the mercury in the tube above a fixed point in the cistern may be observed.

To read the barometer the screw C is turned until the surface of the mercury in the cistern comes to the fixed point. The scale will then indicate the difference in the mercury levels in the tube and the cistern.

Fig. 48.

[112]

2. Siphon Barometer.

Another form of barometer is shown in Fig. 49. The column of mercury in the closed branch A is sustained by the pressure of the air upon the surface of the mercury in the open one B.

The upper scale gives the height of the mercury in the closed branch above a fixed point, and the lower scale the distance of the mercury in the open branch below the same fixed point. The sum of the two readings is the height of the barometer column.

1. Devise an experiment to show that the column of mercury in the tube of a barometer is sustained by the pressure of the atmosphere.

2. What effect upon the height of the barometer would the presence of a piece of iron floating upon the surface of the mercury in the tube have? Explain.

3. At what point in the barometer tube will the pressure of the atmosphere be most likely to crush the tube? Explain.

4. A barometer has some air in the tube above the mercury. How would its reading be affected (1) by taking it down into a coal mine, (2) by raising the temperature of the room in which it is? Explain.

5. Which would be the more suitable for an accurate barometer, a tube of fine bore or one of wide bore? Explain.

Fig. 49.

6. Explain why a barometer falls when carried up a mountain.

8

3. **Find the Pressure of the Atmosphere on a Surface the Area of which is** a**, when the Height of a Barometer Using a Liquid of Density** ρ **is** h.

The pressure of the atmosphere on the surface equals the weight of a column of the liquid used in the barometer whose base is of area a and whose height is h.

But the weight of the liquid

$$= \text{its volume} \times \text{its density}$$

$$= (a \times h) \quad \times \rho.$$

Hence the atmospheric pressure on the surface

$$= ah\rho.$$

EXERCISE XXI.

Note.—In the following questions the density of mercury is to be taken as 13.6 grams per cubic centimetre, or as 13600 oz. per cubic foot.

1. Find the atmospheric pressure per sq. inch when the mercury barometer stands at 30 inches.

2. Find the pressure of the atmosphere on a square centimetre when the mercury barometer stands at 76 cm.

3. Three barometers are constructed to use liquids whose specific gravities are respectively 7.2, 2.9, and 11.8. Find the atmospheric pressure on a square inch (1) when the first barometer stands at 4.8 ft., (2) when the second stands at 11.52 ft., (3) when the third stands at 5.76 cm.

4. Three barometers are constructed to use liquids whose specific gravities are respectively 13.6, 5.17, and 2.06. Find the atmospheric pressure on 1 sq. cm., (1) when the first barometer stands at 70 cm., (2) when the second stands at 2 metres, (3) when the third stands at 5 metres.

5. If in ascending a mountain the barometer falls from 30 in. to 20 in., find the decrease in the atmospheric pressure on an area of 10 sq. ft.

6. If the atmospheric pressure is 15 pounds per sq. in., find the heights of the columns in barometers constructed to use liquids whose specific gravities are respectively 1.44, 3.6, and 4.8.

7. The atmospheric pressure is 1033.6 grams per. sq. centimetre. Find the heights of the columns in barometers constructed to use liquids whose specific gravities are respectively 13.6, 6.8 and 3.4.

8. Find the heights of the columns in barometers constructed to use liquids whose specific gravities are respectively 1.5, 17, and 5 when the mercury barometer stands at 30 in.

9. The mercury barometer stands at 76 cm. What is the height of a water barometer?

10. A mercury barometer stands at 28.8 in. Find the sp. gr. of nitric acid, if a column of it 22.72 ft. in height can be supported by the atmospheric pressure.

. 11. Glycerine rises in a barometer tube to a height of 26 ft. when mercury stands at 30 in. What is the density of glycerine ?

12. At the surface of a lake the barometer stands at 30 in. What will be the reading of the barometer when it is sunk in the water to a depth of 100 ft.?

13. At what depth below the surface of a lake will the barometer stand at 80 in., if at the surface it stands at 30 in.?

14. A mercurial barometer is sunk in a vessel of water, and reads 32.18 in. when its free surface is 3 ft. below the surface of the water, while outside the reading is 29.3 in. Compare the density of mercury with that of water.

15. Solve the following questions in Exercise xvii., taking the atmospheric pressure into account :—(a) Nos. 1, 8, 13, 18, if the barometer stands at 30 in. (b) Nos. 2 and 28, if the barometer stands at 76 cm.

16. Into the upper surface of a closed vessel is let perpendicularly a straight tube. The vessel is completely filled with water, and the

tube is also filled to a height of 11 ft. 4 in. Find the upward pressure of the water on 1 sq. ft. in the upper surface of the vessel, if the height of the water barometer is 34 ft.

17. Water floats on mercury to a depth of 20 ft.; at what depth below the surface of the mercury will the pressure on an area of 1 sq. ft. be 20266 pounds, when the barometer stands at 30 in.?

II.—Air Pump.

1. Construction.

Fig. 50 shows one of the most common forms of the air pump. A cylindrical barrel AB is connected by means of a pipe C with a receiver R, from which the air

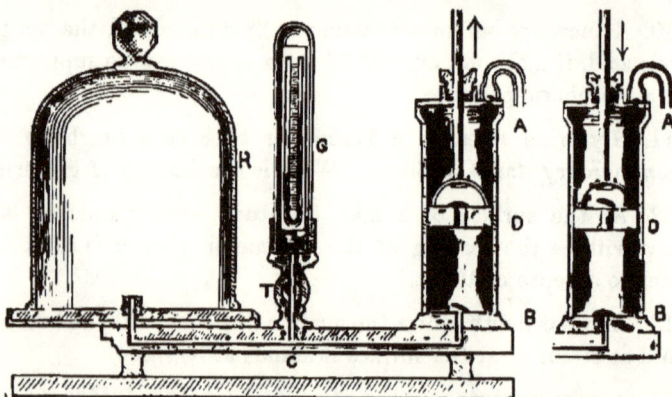

FIG. 50.   50 i

is to be removed. A piston D, in which there is a valve opening upward, is worked in the barrel by a rod which passes through the air-tight collar at the top of the barrel. At A and B, the ends of the barrel, are valves opening upward. A gauge G for testing the extent of the exhaustion is sometimes connected with the tube C by means of a tap T.

2. Action.

Suppose D at its lowest position. As it ascends the compression of the air in AD closes the valve in the piston (Fig. 50) and opens the valve A, and the enclosed air escapes, while a part of the air in the receiver R flows through the valve B and occupies the vacuum formed below D. When the piston begins to descend, the valves A and B are closed (Fig. 50*a*) and the air in DB flows up through the valve in D. Thus at each double stroke of the piston a fraction of the air is removed from the receiver.

1. Of what use is the valve A ?

2. What causes it to close when the piston descends ?

3. What causes the valve in D to open and the valve B to shut when the piston descends ?

3. To Determine the Pressure and the Density of the Air in the Receiver after n Strokes of the Piston.

Let V and v denote the volumes of the receiver and the barrel respectively, P the pressure of the air at first, $P_1, P_2, P_3 \ldots P_n$ the pressures of the air after 1, 2, 3 $\ldots$ n strokes of the piston.

When the piston is first raised, the air in the receiver expands and occupies both the receiver and the barrel.

Thus a volume V of air becomes $V + v$.
But by Mariotte's Law the pressure of the air varies inversely as volume. (Page 126, Part I.)

Therefore $\dfrac{P}{P_1} = \dfrac{V + v}{V}$

or $\qquad P_1 = \dfrac{V}{V + v} P$

Similarly $\qquad P_2 = \dfrac{V}{V + v} P_1$

$$= \left(\dfrac{V}{V + v} \right) \left(\dfrac{V}{V + v} \right) P = \left(\dfrac{V}{V + v} \right)^2 P$$

And so on.

Hence finally $\quad P_n = \left(\dfrac{V}{V + v} \right)^n P.$

The density of the air varies directly as the pressure, hence if ρ denotes the density of the air at first, and $\rho_1\ \rho_2\ \rho_3\ \cdots$ ρ_n, the densities after 1, 2, 3 . . . n strokes

$$\rho_n = \left(\dfrac{V}{V + v} \right)^n \rho.$$

Why cannot all the air be removed from the receiver by means of the air pump?

———

EXERCISE XXII.

1. The capacity of the receiver of an air pump is five times that of the barrel. Compare the density of the air after the fifth stroke with the density at first.

2. If the capacity of the receiver of an air pump is four times that of the barrel, and the initial density of the air is 1, find the density after (1) the 3rd stroke, (2) 5th stroke, (3) 8th stroke of the piston.

3. The capacity of the receiver of an air pump is seven times that of the barrel. After how many strokes will the density of the air in the receiver be to the initial density as 343:512.

4. The volume of the barrel of an air pump is 24 cu. in., when the volume of the receiver is 72 cu. in. The expansive force of the air in the receiver supports a column of mercury 30 inches in height. Find the height of the column of mercury supported after (1) the 5th, (2) the 15th, (3) the 20th stroke.

5. The volume of the receiver of an air pump is ten times that of the barrel, and a barometer placed under the receiver stands at first at 732.05 mm.; but after a certain number of strokes, it stands at 500 mm. Find the number of strokes.

6. Find the ratio of the volume of the receiver to that of the barrel of an air pump, if at the end of the 3rd stroke the density of the air in the receiver is to the original density as 1331:1728.

7. A barometer under the receiver of an air pump stands at 768 mm., but after four strokes of the piston it stands at 243 mm. Find the ratio of the volume of the receiver to that of the barrel.

8. If the receiver of an air pump holds 100 gm. of air at the ordinary pressure and the barrel holds 10 gm., what will be the weight of the air in the receiver after the third stroke ?

9. In a condenser, or air pump for compressing air into a vessel, the valves open downward as shown in Fig. 51. Explain its action, and show how to find the density and the pressure of the air within the vessel after n strokes.

· 10. The volume of the receiver of a condenser is twelve times that of the barrel. Compare the density of the air after the sixth stroke with its original density.

11. The volume of the barrel of a condenser is 10 cu. in., and that of the pneumatic tire of a bicycle into which the air is forced is 100 cu. in. If there is a weak point in the tire just capable of sustaining a pressure of 15 pounds to the square in., after how many strokes of the piston will the tire burst, the original pressure of the air being 15 pounds to the square inch ?

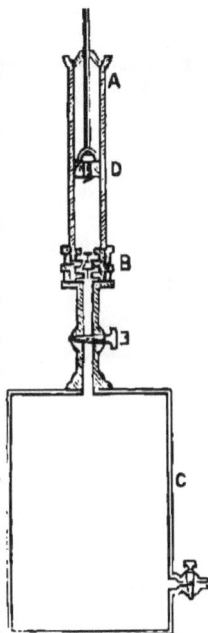

Fig. 51.

III.—Common Pump.

This pump is used for drawing water from a well or cistern.

1. Construction.

The construction is shown in Fig. 52. A cylindrical barrel AB is joined to one end of a suction-pipe BC, the other end of which is placed in the water to be drawn. A piston D, in which there is a valve opening upward, is worked in the barrel by means of a piston-rod. At B there is a valve opening upward. At G a hole is made in the barrel and a spout is inserted.

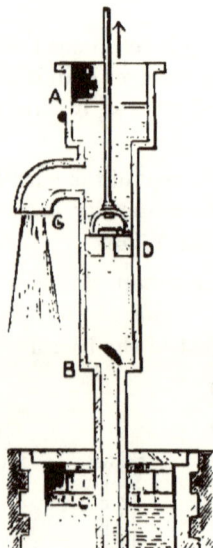

FIG. 52.

2. Action.

Suppose the piston D at its lowest position. As it ascends the valve in D is closed, while the air in the suction-pipe expands, opens the valve B (Fig. 52), and a part of it occupies the vacuum formed below D. Now, since the air which was in the suction-pipe BC occupies a larger space DC, its pressure on the water in the tube will be less than that of the external air upon the surface of the water at C; consequently this external pressure will cause the water to rise until equilibrium is restored, that is, until the expansive force of the air below D, together with the pressure at C due to the weight of the column

FIG. 52a.

of water above it, equals the external pressure of the atmosphere at C. (See Exp. 7, page 119, Pt. I.) When the piston descends, the valve B is closed and the valve in D opened by the compression of the air in DB (Fig. 52*a*), and the air in the lower part of the barrel passes through the valve in B, while the water remains at the same level in the suction-pipe.

At each subsequent stroke the water rises still further in the suction-pipe, and at length forces open the valve B, enters the barrel, passes up through the valve in the piston when it is descending, and is carried forward and thrown out at the spout G when the piston is re-ascending.

1. What is the greatest length which BC can have ? Why ?

2. How could a common pump be used to lift water from a very deep well ?

3. Why does the water stand in the suction-pipe when the piston is not being worked ?

4. A small hole is frequently made in the suction-pipe of a pump to prevent the freezing of the water in the pipe. If such a hole is bored, how high will the water stand in the pipe when the piston is not being worked ? Why ?

5. The top of a well is covered and sealed air-tight. How will this affect the working of the pump (1) when the well is filled with water, (2) when it is partially filled ? Explain.

IV.—Force Pump.

1. Construction.

The construction is shown in Fig. 53. A cylindrical barrel AB is joined to one end of a suction-pipe BC, the other end of which is placed in the water to be drawn. A solid piston is worked in the barrel by a piston-rod. At D a hole is made in the barrel and a pipe E inserted. At B and D are valves opening outward.

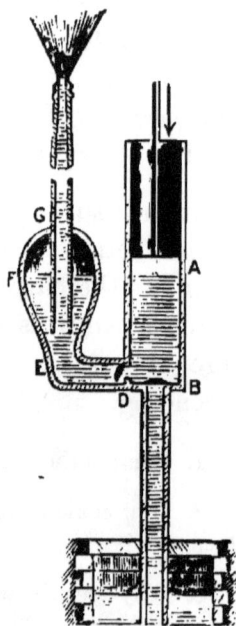

2. Action.

Suppose the piston at its lowest position. As it ascends the air in BC expands, opens the valve B and a part of it occupies the vacuum formed below the piston, the water

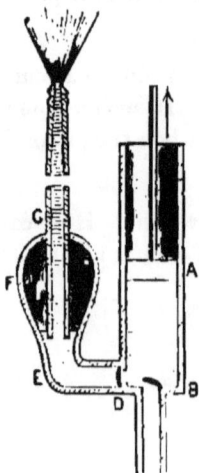

Fig. 53.

being forced up into the pipe BC by the pressure of the air upon the surface of the water in the well. When the piston descends, the air in AB being compressed closes the valve B and flows out through the valve D. At each subsequent stroke of the piston the water rises higher in the suction-pipe, and at length flows into the barrel, when by the descent of the piston the valve B is closed and the water forced through the valve D (Fig. 53) into the tube E. If

Fig. 53a.

the pipe G is connected directly with the pipe E, the stream flowing through it will be intermittent, as it is only on the descent of the piston that the water is forced through D. To produce a continuous stream an interruption is made in the tube which is surrounded by a strong air-tight vessel F. When the water is pumped into this vessel and rises above the open end of the tube G, which is somewhat smaller than the pipe E, the air in the vessel is compressed, and by its expansive force presses the water through the pipe G even when the piston is ascending, thus producing a continuous stream (Fig. 53a).

3. Tension of the Piston-rod in a Common Pump and in a Force Pump.

The tension of the piston-rod is the difference between the pressures on the upper and the lower surfaces of the piston.

If Π is the measure of the atmospheric pressure, h_1 the height of the water above the piston, and h_2 the height of the column of water between the piston and the surface of water in the well, the pressure on the upper surface of the piston is $\Pi + h_1$, while that on the lower surface is $\Pi - h_2$, since the weight of the column h_2 counterbalances a part of the atmospheric pressure. Therefore the tension of the piston-rod is

$$(\Pi + h_1) - (\Pi - h_2) = h_1 + h_2.$$

Hence

The force necessary to raise the piston in the common pump is equal to the weight of a column of water whose base is the surface of the piston and whose

height is the vertical distance between the levels of the water in the pump and the well.

What is the tension of the piston-rod of the force pump on (1) the up-stroke, (2) the down-stroke ?

EXERCISE XXIII.

1. What is the greatest height to which water can be raised by a common pump when the mercury barometer stands at 76 cm., the sp. gr. of mercury being 13.6 ?

2. How high can sulphuric acid be raised by a common pump when the mercury barometer stands at 27 in., the sp. gr. of sulphuric acid being 1.8 and that of mercury being 13.6 ?

3. The area of a piston of a common pump is 4 sq. in. Find the tension of the piston-rod when the water stands in the suction-pipe at a height of 12 feet above the water in the well.

4. If the spout of a common pump is 21 feet above the surface of the water in a well, and the diameter of the piston is 4 in., find the tension of the piston-rod when the pump is full of water.

5. The diameter of the piston of a common pump is 6 cm., and the tension of the piston-rod is 13,750 grams when the pump is full of water. Find the distance from the spout to the surface of the water in the well.

6. The tension of the piston-rod of a common pump is 300 pounds when the water in the suction-pipe is 16 feet above the surface of the water in the well. Find the area of the piston.

V.—Bramah's Press.

1. Construction.

The construction of Bramah's Press is shown in Fig. 54.

It consists of a force pump A, the tube of which opens into a cylindrical vessel B with very strong, thick sides.

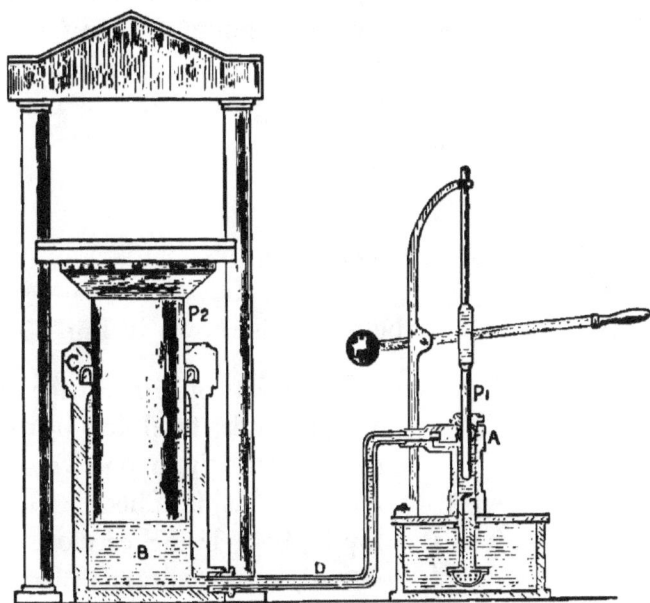

FIG. 54.

In this cylinder there is a large piston or ram P_2 working water-tight in a collar C. A plate to hold the bodies to be pressed is attached to the upper end of the ram. Above this plate is a stationary one supported by the frame work of the machine. The piston of the pump is worked by a lever.

2. Action.

When the water is pumped into the cylinder the ram P_2 is forced upward and the body is pressed between the two plates. Since the pressure is transmitted equally in all directions by the water (Art. 3, page 81), the pressure exerted upon the ram P_2 will be as many times that applied to the piston-rod of the pump as the area of a cross-section of the ram P_2 is that of the area of a cross-section of the piston P_1 of the pump. Therefore, by decreasing the size of the piston of the pump, and increasing that of the ram very great pressure may be developed by the machine.

VI.—Siphon.

1. Construction.

The construction of the siphon is shown in Fig. 55.

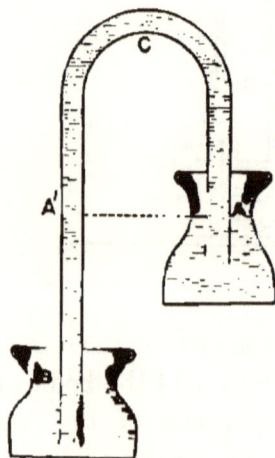

It consists of a bent tube open at both ends, used for transferring a liquid from one vessel to another. It is filled with the liquid, both branches closed, inverted, and one branch placed in the liquid to be transferred. The end of the other branch must be below the surface of the liquid in the vessel from which the liquid is to be withdrawn.

Fig. 55.

When the ends are unstopped the liquid will run in a continuous stream through the tube.

2. Explanation.

The pressure at A tending to move the liquid in the siphon in the direction AC

= the atmospheric pressure — the pressure due to the weight of the liquid in AC

and the pressure at B tending to move the liquid in the siphon in the direction BC

= the atmospheric pressure — the pressure due to the weight of the liquid in BC.

But since the atmospheric pressure is the same in both cases, and the pressure due to the weight of the liquid in AC is less than that due to the weight of the liquid in BC, the force tending to move the liquid in the direction AC is greater than the force tending to move it in the direction BC; consequently a flow takes place in the direction ACB. This will continue until the vessel from which the liquid flows is empty, or the liquid comes to the same level in each vessel.

EXERCISE XXIV.

1. Is the pressure due to the weight of the water in AC equal to the actual weight of the water in AC? If not, to what is it equal?

2. Upon what does the rapidity of flow in the siphon depend?

3. Will a siphon work in a vacuum? Explain.

4. Upon what does the limit of the height to which a liquid can be raised in a siphon depend?

5. How high can water be raised in a siphon when the mercury barometer stands at 30 in., the sp. gr. of mercury being 13.6?

6. How high can sulphuric acid be raised in a siphon when the mercury barometer stands at 27 in., the sp. gr. of the acid being 1.8 and that of the mercury 13.6?

7. Find the greatest height over which a liquid of density ρ_1 can be carried when the height of the barometer is h, the density of the liquid used in the barometer being ρ.

8. What would be the effect when the siphon is working of making a hole in it (Fig. 55), (1) at C, (2) between A and C, (3) at A^1, (4) between A^1 and C, (5) between A^1 and B?

Fig. 56.

9. Will any change in the action of a siphon be coincident with a fall in the barometer? Explain.

10. Make a piece of apparatus similar to that shown in Fig. 56, by cutting the bottom off a bottle, bending a glass tube and inserting it into a perforated cork placed in the bottle. Let water from a tap run slowly into the bottle. What takes place? Explain.

11. Natural reservoirs are sometimes found in the earth, from which the water can run by natural siphons faster than it flows

Fig. 57.

into them from above (Fig. 57). Explain why the discharge through the siphons is intermittent.

CHAPTER XIII.

I.—Vibratory Motion.

1. Vibration.

Experiment 1.

Suspend a weight by means of a wire or cord. Draw the weight aside, let it go and observe its motion. (Fig. 58.)

1. Describe the changes in velocity which take place.

2. Is the number of times which it moves to-and-fro the same during equal intervals of time ?

When the motion of a body, like that of the suspended weight, is alternate in direction, it is said to be **oscil latory**, or **vibratory**.

Fig. 58.

When the number of vibrations during any given interval of time is constant, the motion is said to be **periodic.**

The time required to perform a complete vibration is called the **period of vibration.**

The number indicating the number of vibrations in a unit of time is called the **vibration-number**, or **vibration-frequency.**

9

If τ denotes the period of vibration, and n the vibra-tion-number

$$\tau = \frac{1}{n}$$

or $n = \frac{1}{\tau}$

The extent of the excursion of the vibrating body on either side of the middle point, or point of rest, is called the **amplitude** of the vibration.

What change takes place in the amplitude of the vibration, as the weight (Exp. 1) moves to-and-fro ?

2. Direction of Vibration.

Fig. 59.

Experiment 2.

Fasten a steel spring in a small vice, as shown in Fig. 59, draw it aside and let it go. Observe its motion.

When the direction of the motion of the vibrating body is at right angles to its length, the vibrations are said to be **transverse.**

Experiment 3.

Attach a weight to the end of a suspended coil spring, as shown in Fig. 60, pull it down a little way and let it go.

1. Describe its motion.

2. Is it periodic?

When, as in this case, the body vibrates length-wise, the vibration is said to be **longitudinal.**

Experiment 4.

Attach a pointer to a weight, and suspend it, by means of a wire, over a graduated circle drawn on paper, as shown in Fig. 61. Twist it around and let it go.

· 1. Describe the motion of the weight.

2. Is it periodic ?

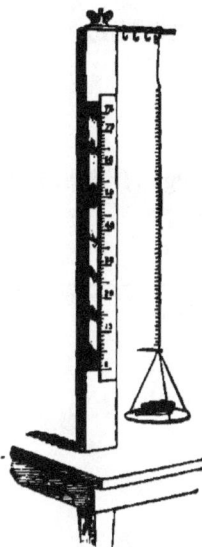

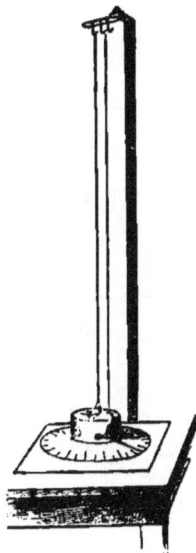

FIG. 60.   FIG. 61.

When a body vibrates by twisting in alternate directions, the vibration is said to be **torsional**.

1. What is the direction of the vibration in each of the following cases ?

(1) When a pendulum vibrates.

(2) When a violin-string is set vibrating.

(3) When the body of a carriage moves up and down on account of the elasticity of the springs.

(4) When the hair-spring of a watch vibrates.

(5) When a flag waves in the wind.

2. Give additional examples of bodies vibrating :—

(1) Transversely.

(2) Longitudinally.

(3) Torsionally.

II.—Sound Caused by Vibration.

3. Vibration of Strings.

Experiment 1.

Stretch a string tightly between two pegs (Fig. 62). Pluck it.

Fig. 62.

Is the sound produced accompanied by vibrations of the string?
How do you know?

4. Vibration of Rods.

Experiment 2.

Cut a ball about the size of a pea from a cork, dip it in some spirit varnish, and when dry attach to it a piece of fine silk fibre. Place a short brass rod in a vice as shown in Fig. 63. Draw a violin-bow across the upper end of the rod, and, holding the free end of the silk fibre in the hand, bring the cork ball so that it will just touch the upper end of the rod when it is giving forth a sound.

Fig. 63.

What takes place? Explain the reason.

Experiment 3.

Fasten the middle of a brass rod about two feet long in a vice as shown in Fig. 64. Suspend by two fibres the cork ball used in the last experiment, and bring it so that it will just touch the free end of the rod. Rub the rod lengthwise with a piece of leather upon which powdered resin has been sprinkled.

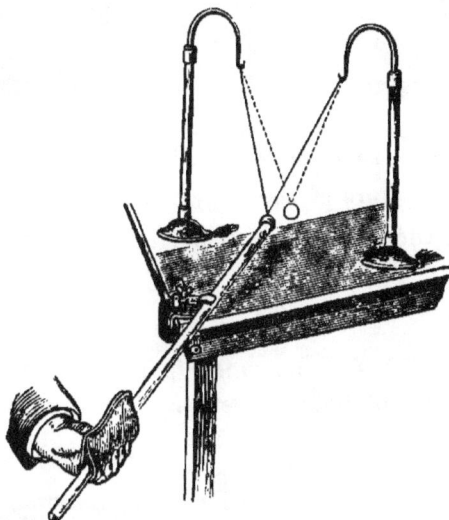

FIG. 64.

1. What evidence have you that the rod is vibrating ?

2. How does its manner of vibration differ from that of the rod in the last experiment ?

5. Vibration of a Tuning-Fork.

Experiment 4.

Place a match on the ring of a retort stand (Fig. 65). Strike the end of a prong of a tuning-fork a sharp blow on a piece of rubber or thick paper folded over the edge of the table. Now touch one of the prongs of the fork to the middle of the match.

What takes place ? Explain the reason.

Strike the fork again and immediately place the prongs so as just to touch the surface of some water.   (Fig. 66.)

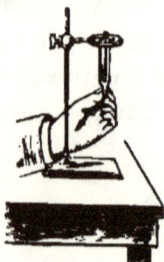

Fig. 65.

Fig. 66.

1. What evidence have you that the fork is in motion ?

2. What is the nature of the motion of the fork ?

To answer the last question perform the following two experiments.

Experiment 5.

Arrange apparatus as shown in Fig. 67.   The block is made by nailing pieces of board together.   The tuning-fork is so

Fig. 67.

placed that the point of a fine style attached to the lower side of one of the prongs just touches the upper smoked surface of a piece of glass, placed on the table below it.   A straight strip of board is tacked to the table to serve as a guide for the glass. Set the fork in motion by drawing a violin-bow across one of the prongs, and slide the glass quickly along the guide, moving it at a uniform rate.

1. Hold the glass up to the light, and describe the form of the tracing.　2. How was the prong of the fork moving?

Experiment 6.

Insert the fork into the centre of one of the ends of the block, and place it in an inclined position, as shown in Fig. 68. Set the fork in motion as before, and, holding the end of the fibre in your hand, bring the ball used in Experiment 2 so that it will touch a prong of the fork just above the crotch. Slowly raise the ball, keeping it alongside the fork.

1. Describe the motion of the ball.

2. When is its displacement by the fork the greatest? When the least?

3. How does the prong of the fork move?

FIG. 68.

Withdraw the fork from the block, strike one of the prongs a sharp blow, and bring the ball against the end of the handle.

1. What takes place?

2. How does the handle of the fork vibrate?

These experiments show that the prongs of the fork oscillate about stationary points PP (Fig. 69) near the crotch, and that the handle vibrates longitudinally.

FIG. 69.

Points of rest in a vibrating body are called **nodal points** or **nodes**.

6. Vibration of Plates.

Experiment 7.

Take a brass plate about 3 millimetres in thickness and about 20 or 25 centimetres square, and hold it in a horizontal position by a suitable clamp at the centre (Fig. 70). Scatter fine sand over the plate, and damp the middle point of one of the edges by touching it with the finger-nail. Now draw a violin-bow across the edge near one of its corners, causing the plate to give a clear strong sound. Observe the movement of the sand.

Fig. 70.

If the experiment is properly performed, the sand will be tossed about and gather in lines, as shown in Fig. 71.

Repeat the experiment, damping a corner of the plate, and drawing the bow across the middle of one of the edges.

Make a drawing showing the position taken by the sand.

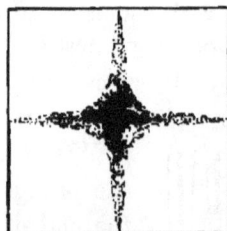

Fig. 71.

Experiment 8.

Repeat Experiment 7, using a circular plate, damping a point on the circumference and drawing the bow at a point 45° from it.

Vary the experiment by bowing the plate at other points.

Make a drawing of the position taken by the sand in each case.

The plates vibrate in divisions, or sections, separated from one another by nodal lines, or lines which remain nearly at rest. While one division moves upward, the adjoining one moves downward. The sand is thus tossed about and is soon thrown to the points of least motion.

7. Vibration of Air-Columns.

Experiment 9.

Insert a whistle* through a perforated cork placed in the end of a glass tube about three times as long as the whistle and 2 centimetres in diameter. Place in the tube some fine precipitated silica, or some cork dust made by filing a cork. Cork the open end of the tube (Fig. 72). Distribute the powder throughout the tube and blow on the whistle. Observe the motion of the powder.

FIG. 72.

1. What evidence have you that the air within the tube is vibrating ?

2. Are there any nodes, or points of rest, in the air column within the tube ?

To show that the movement of the powder is not caused by the vibration of the glass or whistle, grasp these tightly in your hands and repeat the experiment.

The air driven through the mouth of the whistle strikes against the sharp edges of the lateral opening of the mouth-piece, thus causing vibrations of the air within the whistle and the tube. The vibrations are made visible by the vibrations of the light powder moving with the air.

* A common tin or wooden flageolet cut off at the orifice nearest the mouth-piece will answer.

8. Origin of Sound.

The preceding experiments in illustration of the vibrations of strings, rods, plates, columns of air, etc., show that in each case when sound was produced it was accompanied by vibrations in some body. The most careful examination of all sounding bodies shows that this is always the case, and that **the sensation of sound has its origin in vibrations of some vibrating body.**

Vibrations which give rise to the sensation of sound are called **sonorous vibrations.**

9. Nature of Sound.

Sound is defined sometimes as a **sensation,** and sometimes as the **external cause of the sensation.**

As a sensation, it is an effect perceived normally by the ear.

As the antecedent cause of a sensation, it is defined as **that special condition of matter in virtue of which incidentally it may affect the organ of hearing.**

The term is generally used in the latter sense in physics.

CHAPTER XIV.

I.—Medium of Transmission.

1. A Material Medium Necessary for Transmission of Sound.

Experiment 1.

Place on the plate of an air-pump a thick tuft of cotton batting, and upon it place an alarm-clock or a bell rung by clock-work or by an electric current (Fig. 73). Cover the whole with a receiver, and, when the bell is ringing, gradually exhaust the air from the receiver.

What change takes place in the sound as the air is exhausted ?

FIG. 73.

When the exhaustion is as complete as is possible to make it, let the air gradually into the receiver.

1. What change now takes place ?

2. What therefore seems necessary to transmit the sound to the ear ?

2. Sound Transmitted by Solids.

Experiment 2.

Repeat the last experiment, removing the batting and placing the sounding body on the plate of the pump.

1. How do your observations in this case differ from those made in the previous experiment ?

2. How can you account for the difference?

[139]

Experiment 3.

Glue the centre of a thin board 6 or 8 inches square to the
end of a pine rod 4 or 5 feet long and one inch square. Screw
a hook into the other end of the rod, hang a watch on the
hook and place your ear close to the board.

Can you hear the ticking of the watch ?

Remove the rod, place the watch at the same distance and
try whether the ticking can be heard through the air.

In which case is the sound transmitted with the greater distinctness?

Experiment 4.

Repeat the last experiment, placing the ear to the board
when another person touches the end of the handle of a
vibrating tuning-fork to the end of the rod, removes it and
again touches it to the rod.

What changes in the distinctness of the sound are observed ?

Experiment 5.

Place a watch on a table, bring your head near the watch
and stop your ears so that the ticking can not be heard. Now
touch your forehead to the watch.

What is observed? Why?

Vary the experiment by touching the watch to your teeth
and to other parts of your head.

3. Sound Transmitted by Liquids.

Experiment 6.

Unscrew a tuning-fork, which has been mounted on a
resonance-box, from the box. Fill a tumbler or glass jar
nearly full of water and place it on the resonance-box. Now

stick the handle of the fork into a large cork, excite the
fork, and place the cork in
the water (Fig. 74). Alternately raise the cork out of
and lower it into the water.

1. What changes in the sound
are observed?

2. What media transmit the
sound when the cork is in the
water?

3. It is said that divers under
water can hear sounds made on shore.   How may this be possible?

FIG. 74.

The above experiments show that **a material medium
is required for the propagation of sound, and that
solids and liquids as well as gases transmit sound.**

II.—Wave-Motions.

Before proceeding to consider the theory of the transmission of sound by elastic bodies, it will be necessary for
the student to become familiar with the phenomena of
wave-motion in general.

4. Transverse Waves.

Let a pebble drop into a body of water at rest, and
observe the motion of the water.

Observe

(1) That a depression is produced at the point where the stone
touches the water.

(2) That this depression travels outward from this point as a
circular trough.

(3) That the depression of the water at the point where the stone dropped is followed by an upward movement of the water at this point, causing a ridge or crest.

(4) That this crest travels after the depression, moving at the same rate in a circle concentric with it.

(5) That this crest is followed by another depression, and the depression by another crest, and so on, thus producing a series of ripples or waves.

Throw a piece of wood on the water, and note that it moves up and down, thus showing that while the waves move outward in a horizontal direction, the particles vibrate vertically.

Experiment 1.

Fasten one end of a light chain about 8 feet long to the ceiling (Fig. 75). By giving the lower end of the chain a number of quick jerks, send a series of pulses along the chain.

1. In what direction are the waves in the chain propagated ?

2. In what direction do the individual links of the chain move?

Waves, like those of the water or the chain, **which are produced by the vibratory motion of particles at right angles to the direction in which the wave is propagated, are called transverse, or crest-and-hollow, waves.**

FIG. 75.

The distance between two successive crests or between two successive hollows is called a **wave-length**.

5. Longitudinal Waves.

Experiment 2.

Make a "wave machine" similar to that illustrated in Fig. 76. The spiral should be 2 metres long and 7 cm. in diameter, and be made of 72 turns of No. 12 copper wire.

The spiral may be made by winding the wire uniformly around a cylindrical rod, and then removing the rod.

The threads by which the spiral is suspended should be 60 cm. long. The frame should be made light and stiff.

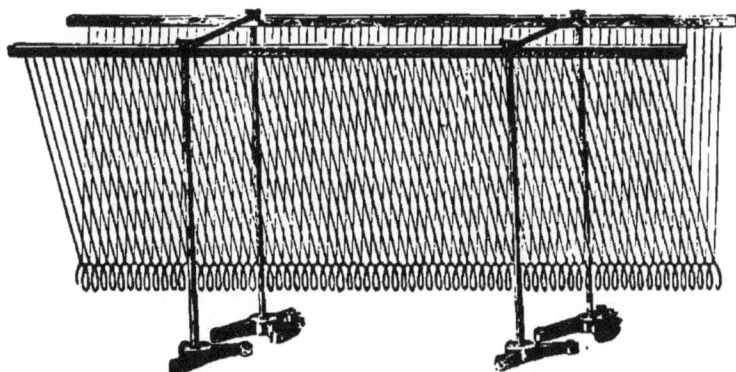

FIG. 76.

Take hold of the end of the spiral and give it a quick jerk outward in the direction of the axis, and then let go. Observe the pulse as it moves along the spiral.

How do the separate coils move?

Take hold of the end of the spiral again, push it inward, let go quickly, and again observe the pulse as it moves along the spiral.

How do the separate coils now move?

Insert the blade of a knife between the coils near one end of the spiral, rake it quickly towards the other end across a few turns of the wire, and observe the motion of the wave along the spiral.

In the first case, when the spiral was jerked outward, the coils appeared to move backward in a direction opposite to that in which the pulse is moving, thus forming what is called **a pulse of rarefaction.** In the second case, when the spiral was pushed inward, the coils appeared to move forward in the direction in which the pulse was moving, forming **a pulse of condensation.**

In the third case a pulse of condensation (Fig. 77 B) is followed by a pulse of rarefaction (Fig. 77 C), and a double pulse, or wave, is formed.

A   B   C

Fig. 77.

It will be observed that while the wave moves along the spiral, each individual turn of wire simply moves backward and forward in the line of direction of the wave.

Waves which are produced by the vibratory motion of particles along the lines in which the waves are propagated are called longitudinal waves.

The wave-length in this case is the distance between two successive centres of condensation or between two successive centres of rarefaction.

III.—Theory of the Transmission of Sound by an Elastic Medium.

We have learned that the sensation of sound has its origin in the vibrations of some vibrating body (Art. 8, page 138). We are now to consider the manner in which sounds are conveyed by the media which transmit them.

Experiment 1.

Wet a piece of linen paper and paste it over the mouth of a tumbler. When the paper has become dry, cut away a part of it, as shown in Fig. 78, making a small opening at first and

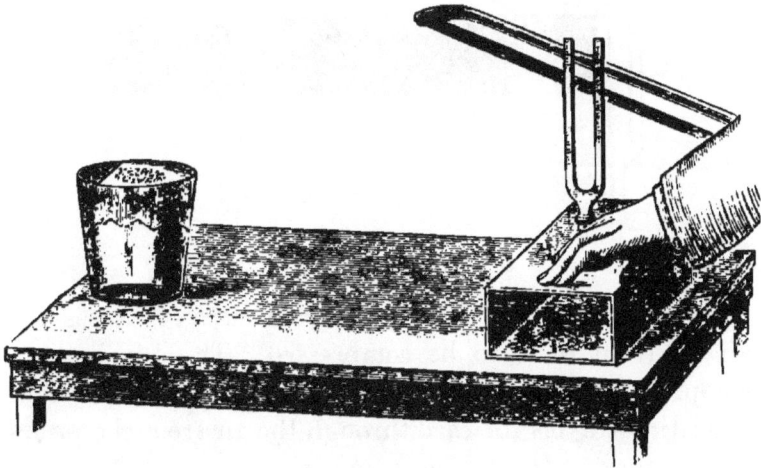

Fig. 78.

gradually increasing its size until the tumbler gives forth a loud sound when a vibrating tuning-fork is held over the opening. Sprinkle fine sand on the paper, mount the fork on a resonance-box, and sound it at a distance from the tumbler.

1. What effect has the sounding of the fork on the sand placed on the paper?

2. What evidence have you that the paper is vibrating?

10

Since the vibrations of the tuning-fork can reach the paper only by passing through the air, the air itself must be in a state of vibration.

The vibrations of the coil spring, Exp. 2, page 143, furnish an illustration of the manner in which the vibrations of a sounding body, for example a tuning-fork, are believed to be transmitted by the air.

As one of the prongs of a vibrating fork swings swiftly forward, it compresses the air immediately in front of it (Fig. 79). This air, on account of its elasticity, resists

FIG. 79.

the compression, and its tendency to expand causes the air in front of it to be compressed. This air in turn compresses that in front of it, and thus a pulse of con-densation travels forward through the air from the prong of the fork. In the meantime the prong of the fork swings backward, and the air behind it is rarefied. A pulse of rarefaction is thus produced, which follows immediately the pulse of condensation. This in turn is followed by a pulse of condensation, which again is followed by one of rarefaction, and so on.

These alternate pulses of condensation and rarefaction constitute a regular series of sound-waves, which pass in succession through the air, and, falling upon the ear, produce the sensation of sound.

The vibration of all other sonorous bodies set up similar waves, which pass outward in every direction from the body, like a series of ever enlarging concentric spherical shells (Fig. 80).

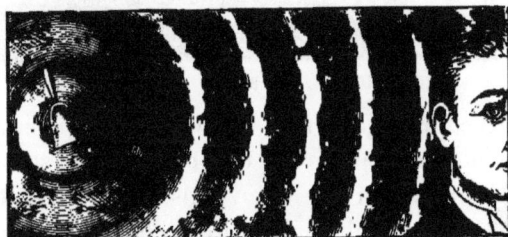

Fig. 80.

The student must be careful not to confound the motion of the wave with the motion of the air particles which constitute it at any instant. **While the wave moves constantly forward, the air particles simply move backward and forward in the direction of the wave.**

Since a single wave is made up of a pulse of condensation and a pulse of rarefaction, **each aerial particle makes one complete vibration while the wave progresses one wave-length.**

To get a clear conception of the propagation of sound-waves and the motion of the individual particles which compose them, repeat the experiment with the coil spring. Also perform the following experiment.

Experiment 2.

Cut a slit AB in a piece of black cardboard as shown in Fig. 81a. Place the slit over the dotted line in Fig. 81b, and draw the book from under it in the direction of the arrow, keeping the slit always at right angles to the side of the page.

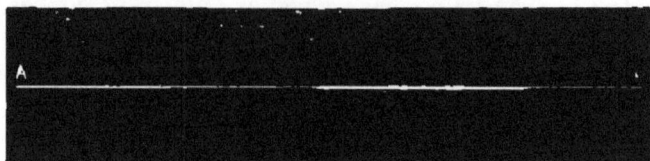

FIG. 81a.

Observe the propagation of the waves of condensation and rarefaction as they appear at one end of the slit and pass along in the direction of the other. Also observe the to-and-fro motion of the individual small white dashes in the direction of the slit.

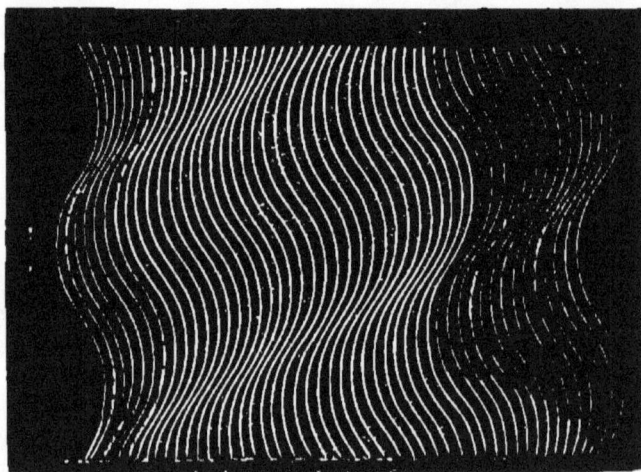

FIG. 81b.

Liquids and gases are believed to transmit sound by waves of condensation and rarefaction in exactly the same way as air.

IV.—Velocity of the Transmission of Sound.

6. Velocity of Sound Dependent on the Elasticity and the Density of the Medium.

The velocity with which sound is transmitted by any medium depends on the elasticity and the density of the medium. The exact law, the discussion of which is beyond the limits of this work, is given by the equation

$$V = \sqrt{\frac{E}{D}}$$

where V denotes the velocity of the sound, E the coefficient of elasticity, and D the density of the medium.

The greater the elasticity and the less the density of the medium, therefore, the more rapidly is sound transmitted by it.

Since the density of a solid or a liquid is greater than that of a gas, sound would naturally travel more slowly through these forms of matter than through gases, were it not that the increase in velocity due to their greater elasticities more than compensates for the decrease in velocity due to increase in density. Hence sound-waves generally travel faster in solids and in liquids than in gases.

QUESTIONS.

1. It is found that when a cannon is placed on ice at a distance from the shore and fired, persons on shore hear two reports. Explain the reason.

2. If a given mass of gas is confined within an enclosed space and then heated, what change will take place in (a) the density, (b) the elasticity of the gas? What effect will these changes have upon the velocity with which the gas transmits sound?

3. If the same gas is heated when it is free to expand, what changes in density and elasticity will take place, and what difference in its conducting power will be observed ? Explain.

4. How does increase in the temperature of the air, the pressure remaining constant, affect the velocity with which sound is transmitted by it ? Explain.

5. If a given mass of gas is confined within an enclosed space and its volume is (1) decreased, (2) increased, while the temperature is kept constant, what change will take place in (a) the density, (b) the elasticity of the gas in each case ? What is the relation between the change in elasticity and the change in density ? (See Part I, page 119.) What changes, if any, will take place in the velocity of sound transmitted by the gas ?

6. What changes in the velocity of sound transmitted by the air accompany (1) a rise in the barometer, (2) a fall in the barometer, the temperature of the air remaining constant ? Explain.

---

7. Determination of the Velocity of Sound in Air.

The velocity of sound in the air may be approximately determined in the following manner. The distance between two stations is measured. A gun is fired at one station and the interval of time elapsed between the seeing of the flash and the hearing of the report at the other station is observed. The distance between the stations and the time taken by the sound to travel between them being known, the velocity of sound in the air can be determined. It is assumed that the time taken by the light to pass from one station to the other is so short that it may be neglected.

To allow for the action of the wind, the firing should be done at alternate stations, and the average of several results taken.

The above method will at best give but an approximate result. That devised by Regnault gives a much more accurate determination.

A pistol is fired at one end of a long tube of known length, and an electrical recording apparatus registers automatically the time that elapses between the pulling of the trigger of the pistol and the making of a mark by a pointer attached to a membrane, which, placed at the other end of the tube, vibrates when the sound-pulse reaches it. Data are thus furnished for calculating the velocity of the sound transmitted by the air in the tube.

The velocity of sound in any other gas may be determined in the same manner with this apparatus. The air is exhausted and the tube filled with the gas.

When the temperature is 0°C, the velocity of sound in the air is about 1090 feet per second, and the velocity increases about two feet per second for each increase in one degree centigrade in temperature.

8. Determination of the Velocity of Sound in a Liquid and in a Solid.

Since the coefficients of elasticity and the densities of liquids and solids can be determined experimentally, the velocity of sound in these forms of matter can be determined theoretically from the equation

$$V = \sqrt{\frac{E}{D}}$$

The results are found to agree closely with experimental determinations when these are possible.

The velocity of sound in water is about $4\frac{1}{2}$ times its velocity in air.

9. Relation among Velocity, Vibration-Number and Wave-Length.

The particles composing a sound-wave make one complete vibration while the sound-wave travels one wave-length; therefore, if n denotes the number of vibrations in a unit of time, and λ the wave-length of the sound-wave, the sound travels a distance of $n\lambda$ in a unit of time, or

$$v = n\lambda$$

where v denotes the velocity of the sound.

Hence

The velocity = the vibration-number × wave-length.

. This relation gives a method of determining the velocity in air when the vibration-number and the wave-length are known. (See Art. 9, page 163, and Art. 2, page 187.)

QUESTIONS.

1. Two stations are 61045 feet apart, and the report of a gun fired at one station is heard at the other 54.6 seconds after the flash is seen. What is the velocity of the sound in the air?

Fig. 82.

2. Two boats were stationed on Lake Geneva, 51700 feet apart. One boat was supplied with a bell B placed under water (Fig. 82).

and so arranged as to be struck by a hammer H at the same instant that a torch M turns over and lights the gunpowder P. The sound of the bell was heard at the other boat, by means of trumpet T placed in the water, just 11 seconds after the flash of the gunpowder was seen. What was the velocity of sound in water?

3. A report of a cannon is heard 6 seconds after the flash is seen. If the temperature of the air is 15C., what is the distance of the observer from the cannon?

4. A tube, 1000 feet long, is filled with oxygen gas. Find how quickly the report of a pistol fired at one end of the tube will pass to the other, if the density of oxygen is 16 times that of hydrogen, and the velocity of sound in hydrogen is 4200 feet per second, when the pressure of the hydrogen is the same as that of the oxygen.

5. In which would you expect sound to travel the more quickly, (1) water or alcohol, (2) water or mercury? Give reasons for your answer.

6. How do you account for the fact that the time required for a sound to travel a certain distance differs from day to day?

7. If all the soldiers in a long column keep time to the music of a band, will they step together?

8. A number of soldiers are drawn up in the form of a circle, and each man fires his gun at the instant a signal is given by a man placed at the centre of the circle. Will the sound appear as a single report to any of the men? Explain.

9. If a tuning-fork which vibrates 256 times per second sets up in the air sound-waves the wave-length of which is 52 inches, what is the velocity of sound in the air?

10. It is observed that the velocity of a sound produced by a tuning-fork whose vibration-number is 435, is 1100 feet per second, what is the wave-length of the sound-waves?

CHAPTER XV.

I.—Intensity of Sound.

1. Intensity and Amplitude of Vibration.

Experiment 1.

Repeat Exp. 1, page 132. Observe the string, and note the changes in the amplitude of its vibrations.

What change takes place in the amplitude of the vibration of the string as the sound grows weaker?

Experiment 2.

Repeat Exp. 5, page 134. Observe the tracing on the smoked glass.

What evidence have you that the intensity of the sound increases with the amplitude of vibration of the sonorous body?

The intensity of a sound-wave is measured by the energy of the vibrating particles. When the vibration-number remains constant, the velocity varies as the amplitude of vibration; for example, if the vibrating particles have twice as far to swing in the same time, the velocity must be doubled: but the energy varies as the square of the velocity (Part I, page 78), therefore the **intensity of a sound-wave varies as the square of the amplitude of vibration.**

[154]

2. Intensity and Density of the Medium.

Experiment 3.

Repeat Exp. 1, page 139.

1. What change takes place in the density of the air in the receiver as the exhaustion proceeds ?

2. What change in intensity of the sound accompanies this change in density ?

The intensity of sound-waves increases with the density of the medium in which they originate.

If the intensity of the sound-wave is measured by the energy of vibrating particles, why should its intensity be affected by changes in the density of the medium in which it originates ? (See Part I, page 75.)

3. Intensity and Distance.

It is a matter of common experience that the intensity of a sound decreases with an increase in the distance from the point at which it originates. The exact law will be learned from the following considerations.

As we have seen (page 147), a sound-wave is spherical in form.

When the radius is unity, the surface is 4π.

"	"	2	"	16π.
"	"	3	"	36π.
"	"	etc.	"	etc.

But each of the surfaces 4π, 16π, 36π, etc., receives the same amount of energy, therefore the energies received by a unit surface are proportional to $\frac{1}{4\pi}$, $\frac{1}{16\pi}$, $\frac{1}{36\pi}$, etc., or to 1, $\frac{1}{4}$, $\frac{1}{9}$, etc., when the distances are in the proportion 1, 2, 3, etc.

Hence

The intensity of a sound-wave varies inversely as the square of the distance from its source.

4. Reinforcement of Sound—Consonance.
Experiment 3.

Excite a tuning-fork, observe the loudness of the sound, and press the end of the handle against a table.

What change takes place in the loudness of the sound?

Excite the fork and observe the time the sound continues (1) when it is held in the hand, (2) when the end of the handle is pressed against the table.

1. In which case does the sound continue the longer?

2. What is the cause of the difference?

When the two were in contact, the fork communicated its vibrations to the table, a greater mass of air was set in vibration, and the intensity of the sound was consequently increased.

When the intensity of a sound of a vibrating body is reinforced by the communication of its vibrations to a body of greater surface, thus generating air-waves of greater volume, the effect is called consonance.

Sound may be reinforced also by **resonance**. This will be considered at a later stage. (See page 184.)

QUESTIONS.

1. The workmen engaged in constructing the St. Clair tunnel observed that when working in an atmosphere of compressed air, the tones of ordinary conversation appeared abnormally loud. Explain the reason.

2. Why are sounds produced on high mountains of diminished intensity?

3. Will an increase in the vibration-number cause an increase in the intensity of a sound-wave if the amplitude of vibration remains constant? Give reasons for your answer.

4. A cannon is fired at a point half way up a mountain. Will the report reach two observers with equal intensity, if one observer is at the top of the mountain and the other at the foot? Give reasons for your answer.

5. If two cannons, charged equally, are fired, one at the top of a mountain, and the other at the foot, will the reports come to a person stationed half way up the mountain with equal intensities? Give reasons for your answer.

6. If a small bell is placed over water in a flask, as shown in Fig. 83, the sound of the bell can be distinctly heard when the flask is corked; but if the water is boiled, the lamp removed, and the flask corked while the steam is still issuing from its mouth, the sound of the bell can scarcely be heard after the flask has cooled. Explain the reason.

FIG. 83.

7. If A is 500 yards from a cannon when it is discharged, and B 1000 yards, how many times as loud will the report which reaches A be as that which reaches B?

8. What is the use of the sounding-board of a piano?

9. Why are the strings of a violin mounted on a wooden box made of thin boards?

II.—Reflection of Sound.

Experiment 1.

Cut from a sheet of cardboard a circular disc about 10 inches in diameter, cut out of the disc two sectors, as shown in Fig. 84, and mount it on the spindle of a whirling machine. Let one person rotate the disc, and at the same time blow on a toy trumpet held inclined to the plane of the disc, while another person goes to a distant part of the room and describes the changes which he observes in the sound given out by the trumpet.

Fig. 84.

Is the effect different when the observer stands in different parts of the room?

The effect is due to interrupted reflections of the sound-waves. When the cardboard is in front of the trumpet the sound-waves are thrown back from its surface, and the sound appears to the listener to increase in power. This experiment furnishes a simple illustration of a common phenomenon.

When sound-waves are not obstructed they are propagated, as we have learned in the last chapter, in the form of concentric spheres; but when they encounter an obstacle in any direction they are reflected from the surface upon which they strike, and the direction is consequently changed.

Sound-waves obey the same laws of reflection as ether-waves. (See reflection of light.)

5. Reflection from Concave Surfaces.

Experiment 2.

Find the focus of a concave mirror by letting sunlight fall upon it, and noting the point in front of it at which a match is kindled. Suspend a watch at the focus of the mirror and place another similar mirror facing it at a distance of 6 or 8 feet from it. Connect a glass funnel with your ear by means of a rubber tube, and place the funnel at the focus of the second mirror, with the mouth of the funnel towards the mirror

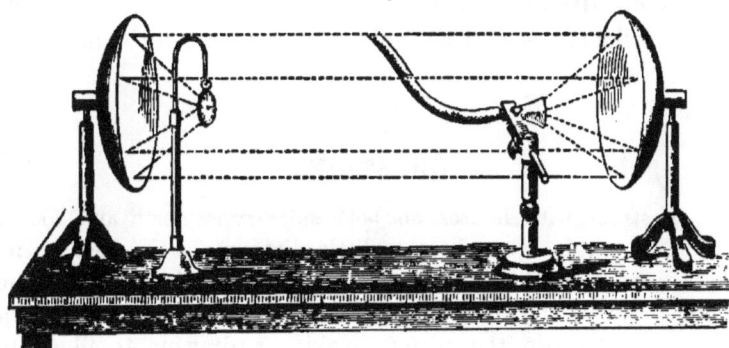

FIG. 85.

(Fig. 85). Place the end of the rubber tube in your ear and listen to the ticking of the watch. Now remove the second mirror and listen again.

1. In which case is the ticking heard with the greater distinctness? Explain the reason.

6. Reflection by Change of Density.

Sound-waves are not only reflected by the surfaces of solids, but also by clouds, gas flames, etc.; and even in passing from a gas of one density to another of a different density they are in part reflected at the surface of separation. Whenever, therefore, sound-waves are propagated by a medium which is not of uniform density, they are more or less speedily dissipated by repeated reflections. Sound usually travels further at night than in the day-time because the air is more homogeneous.

Why is the air more homogeneous at night?

7. Echoes.

Echoes are the result of the reflection of sound-waves. When an interval of time intervenes between a direct and a reflected sound the latter is heard as an echo.

QUESTIONS.

1. A street with houses on both sides runs north and south, and a church is situated at a little distance to the east of it. To a person walking down the eastern side of the street the sound from a bell in the church-tower seems to come from the west. Explain the reason, making a drawing to illustrate your answer.

2. In large buildings and in mountainous regions a succession of echoes is sometimes heard. Explain the reason.

3. A person is walking between two parallel walls which are near together, and hears a prolonged echo of each footstep. Explain how the echo is produced.

4. Why do deaf persons sometimes place their hands behind their ears to catch sounds?

5. How may an echo be made use of for determining approximately the velocity of sound in air?

6. On what conditions may the echo appear louder than the direct sound?

7. In the whispering-gallery in St. Paul's Cathedral the faintest sound is conveyed from one side of the dome to the other, but is not heard at any intermediate point. Explain the reason.

8. A traveller finds himself between two nearly parallel precipitous mountain ranges, which are 2,200 feet apart, while the road on which he is runs through the valley at about 550 feet from one of the sides. On firing his gun he hears three distinct echoes, the second being considerably fainter than the first, while the third is almost as loud as the second. Explain this, and calculate the time elapsed between the hearing of each echo and the firing of the gun, if the velocity of sound in the air of the valley is 1,100 feet per second.

9. Mr. Tyndall, having to lecture in the Senate House of the University of Cambridge, made some experiments as to the loudness of voice necessary to fill the room, and found that a friend, placed at a distant part of the room, could not hear him distinctly. When the audience assembled his voice was plainly heard in all parts of the Senate House. Explain the reason of the change.

10. Why is it that a sound will travel much further through a tube than through the open air?

11. At Carisbrook Castle, in the Isle of Wight, is a well two hundred and ten feet deep and twelve wide. The interior is lined by smooth masonry. When a pin is dropped into the well, it is distinctly heard to strike the water. Explain the reason.

12. How do you account for the fact that the distance which a loud sound, such as the report of a cannon, can be heard varies considerably from day to day?

13. Arctic travellers, separated by more than a mile of frozen water, have conversed with ease. How was this possible?

11

III.—Interference of Sound-Waves.

Experiment 1.

Take two brass U-tubes, A and B, connected by telescoping joints as shown in Fig. 86, short tubes being inserted at C and D. Adjust the tube A so that the distance from the opening C to the opening D will be the same around the tube in the direction CAD as in the direction CBD. Connect D with your ear by means of a piece of rubber tubing, and place a vibrating tuning-fork at the opening C.

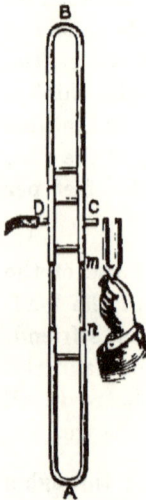

Note that the sound of the fork is heard, the sound-waves reaching the ear through both branches together.

Now draw A out until the intensity of the sound is a minimum.

FIG. 86.   If A is properly adjusted the sound will almost, or altogether, disappear.

This will take place when the sound-waves passing through A reach the point D just one-half a wave length behind those which leave C at the same time and pass through B. Pulses of rarefaction at D are always met by pulses of condensation, and the sound-waves are consequently destroyed.

In this case the distance around the tube in the direction CAD differs from the distance in the direction CBD by one-half a wave-length, that is, the distance *mn* is one-quarter of the length of the wave produced by the fork.

When two sound-waves interfere and thus destroy each other either wholly or partially the effect is known as interference.

8. Determination of Wave-Length.

The instrument described in the experiment above may be used for determining the wave-length of the sound-wave produced by a tuning-fork.

The difference in the lengths of the paths CAD and CBD when complete interference takes place is one-half the wave-length of a sound-wave produced by the fork. That is, if m and n are together when the two paths are equal in length, the length of the sound-wave $= 4\ mn$.

9. Determination of the Velocity of Sound in Air by Determination of Wave-length.

The above furnishes an indirect method of determining the velocity of sound in air. If the vibration-number of the fork is known, and the wave-length determined, the velocity in air can be found from the relation

$$v = n\lambda. \qquad \text{(Art. 9, page 152.)}$$

10. Beats.

Experiment 2.

Take two tuning-forks of the same vibration-number, hold one in each hand and make them sound together.

Note that the sounds blend perfectly.

Load one of the forks by sticking a piece of wax to the end of one of the prongs. Excite the forks and place each on its resonance-box.

Note that there is now no continuous flow of sound, but that the intensity is alternately increased and diminished.

This effect is the result of interference, and is called **beating.**

The loading of the fork causes it to vibrate more slowly and to originate sound-waves which are of greater wave-length than those produced by the other fork. Since the sound-waves proceeding from the forks are equal in velocity and unequal in length, they periodically coincide, the condensation of the one with the condensation

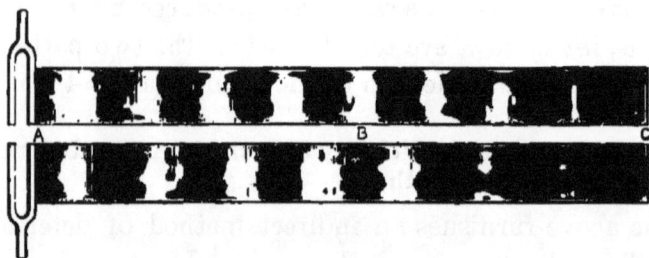

FIG. 87.

of the other (Fig. 87 A), or the rarefaction of the one with rarefaction of the other (Fig. 87 C); and periodically interfere, the condensation of the one coinciding with the rarefaction of the other (Fig. 87 B). Thus alternate reinforcements. and diminutions of sound are produced.

QUESTIONS.

1. Vibrate a tuning-fork, hold it upright near the ear, and slowly rotate it on its axis. What changes take place in the sound as it is turned around once? Explain the reason.

2. If a circular plate is made to vibrate in four sectors, as in Exp. 7, page 136, and if a cone shaped funnel is connected with the ear by a rubber tube, and the other ear is stopped with soft wax, no sound is heard when the centre of the mouth of the cone is placed over the centre of the plate; but if it is moved outward along the middle of a vibrating sector, a sound is heard. Explain

the reasons. Try the experiment. The mouth of the funnel should be about $2\frac{1}{2}$ inches in diameter, if the diameter of the plate is 6 inches.

3. A sounding tuning-fork, mounted on a resonance-box, is carried slowly toward the wall of a room. Why is it that the sound becomes wavy, rising and sinking at regular intervals? .

4. A vibrating fork is placed before the opening C in the tubes (Fig. 86), and the observer at D notes that complete interference takes place when A is drawn out 13 inches. What is the length of the sound-wave which originates with the fork? If the vibration-number of the fork is 256, what is the velocity of sound in air?

5. A fork, whose vibration-number is 256, is made to vibrate before the opening C (Fig. 86), and perfect interference takes place when the tube A is drawn out 33 cm. What is the velocity of sound in the air?

6. When two tuning-forks are beating, show that the number of beats per second is always equal to the difference between the vibration-numbers of the forks.

CHAPTER XVI.

I.—Musical Sounds and Noises.

Musical sounds are the result of a regular succession of vibrations which follow a definite law, and produce an effect agreeable to the ear.

A noise is the result of a combination of vibrations which follow no law, or one so complex that the ear fails to understand or appreciate it.

The distinction then between musical sounds and noises is not absolute, but simply one of degree, depending on the complexity.

Musical sounds differ from one another in

(1) Pitch,

(2) Loudness or Intensity,

(3) Quality or Timbre.

Intensity has already been discussed. Quality will be considered at a later stage (see page 183).

II.—Pitch.

Experiment 1.

Fix on the spindle of a whirling-machine a toothed wheel (Fig. 88). Hold a card lightly against the teeth and rotate the wheel, at first slowly, then more and more rapidly. Observe the sound produced.

Experiment 2.

Have a series of holes punched at equal distances in a circular disc along a circle near its circumference. Mount the disc on the spindle of the whirling-machine (Fig. 89). Insert a piece of glass tubing, the bore of which is about the size of the holes in the disc, into a piece of rubber tubing. Place the glass tube before the ring of holes, rotate the disc, at first slowly, then more and more rapidly, and at the same time force air steadily through the tube.

Fig. 88.   Fig. 89.

1. Describe the changes which take place in the sound produced in each of the above experiments as the velocity of the rotating wheel or disc is increased.

2. What is the cause of the sounds produced in the second experiment?

3. In which case is the frequency of the vibration the greater, when the wheel or disc rotates slowly or quickly?

Pitch is the vibration-frequency of a sound.

A **musical note** is made by a body vibrating a definite number of times per second. The greater the number of vibrations per second, the higher is the pitch of the note.

1. Determination of the Pitch of any Note.

The pitch of any note may be determined by means of an instrument called a siren. Fig. 90 shows the construction of a simple form of this instrument. C is a cylindrical air-chamber upon the upper end of which is mounted a circular rotating disc B, which almost touches the upper surface of the cylinder. The disc is perforated at equal intervals along a circle near its circumference. The upper end of the air-chamber is also perforated, the holes corresponding in number, position, and size with those in the disc above. The holes in both the disc and the end of the chamber are drilled obliquely, those in the disc sloping in one direction and those in the end of the chamber in the opposite. The tube D at the lower end of the chamber is connected with a bellows or blower.

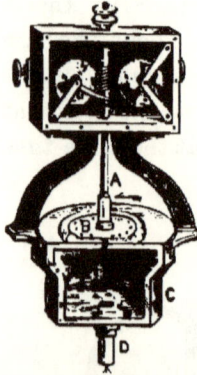

Fig. 90.

When air is forced into the chamber and passes up through the holes, the disc is made to rotate by the pressure of the air against the sides of the holes, the rapidity of rotation depending on the force with which the air is sent into the chamber.

As the disc rotates vibrations will be set up in the external air by the puffs of air which pass out of the chamber when the holes in the disc are opposite those in the end of the chamber. A note will thus be produced, the pitch of which will depend on the rapidity with which the disc is rotated. By controlling the blower any note can be produced at will; and its number of vibrations per second can be determined by reading from the dial of a mechanical recorder attached to the spindle A, on which the disc is mounted, the number of revolutions made in any observed interval, multiplying this by the number of holes in the disc and dividing by the number of seconds in the interval.

Experiment 3.

If your laboratory is supplied with a siren, determine with it the number of vibrations per second of a tuning-fork. Excite the tuning-fork with a violin-bow, and at the same time press air through the siren, gradually increasing the speed of the rotating disc.

When distinct "beats," which indicate that the two notes are nearly alike in pitch, are heard, cautiously increase the speed until the beats disappear and the two notes blend. Now set the clock work of the recorder in motion and keep the disc revolving at a uniform rate for an interval of time, say one-half minute. Read from the dial the number of revolutions, and calculate the vibration-frequency of the fork.

Experiment 4.

If the laboratory is not supplied with a siren, determine approximately the vibration-frequency of the tuning-fork in the following manner :

Take a glass tube about 15 inches long and $\frac{3}{4}$ inches in diameter, and select a cork that will just slide up and down within the tube, touching its sides. Attach a wire to the cork to serve as a handle. Insert the cork into one end of the tube, excite the fork and hold it over the other end of the tube. By means of the wire move the cork up or down until the position of the cork which causes the tube to give out its loudest sound is found. Now place the tube in a horizontal position in a support with its open end close to the disc used in Exp. 2, page 167, and facing the ring of holes (Fig. 91). Hold the tube through which the air is blown on the other side of the disc as shown in the figure, force air through the tube and turn the disc, gradually increasing or decreasing its speed until the velocity at which the tube gives out the loudest sound is found. Continue to revolve the disc at this rate for half a minute, and count the number of turns made by the handle in that time. Multiply the number of turns made by

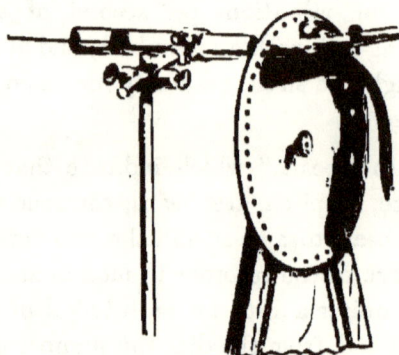

FIG. 91.

the handle by the number of times which the disc revolves for every turn of the handle, and this by the number of holes in circle. Divide the product by 30, the number of seconds. The result will be the number of vibrations per second which the disc sends into the tube ; but, since the tube is sounding its loudest, this is the number of vibrations made by the tuning-fork. The average of several results should be taken.

QUESTIONS.

1. When a sounding body approaches the ear, or recedes from it, the pitch of the tone appears to change. Explain the cause.

2. A person carries a vibrating tuning-fork. Will the pitch of the note appear the same to a person going before him as to a per-

son following him, all three moving at the same rate? Give reasons for your answer.

3. If A carries a vibrating tuning-fork from B to C, will the pitch of the fork appear the same to A, B and C. If not, what will be the difference in their observations?

III.—Musical Scales.

2. Limits of Musical Sounds.

The sense of hearing varies widely in different persons. The ordinary ear is sensitive to vibrations ranging from 16 to 15,000 or 20,000 vibrations per second. The average range of the notes employed in music extends from 40 to 4,000 vibrations per second.

3. Musical Scales.

To produce an effect agreeable to the ear the notes between the limits named above cannot be used arbitrarily or at hazard. When once a note is chosen to begin a piece of music the notes which are to accompany or follow it must be selected according to well defined laws.

In the music of all nations changes in pitch take place by definite intervals, and not by continuous transitions. Music, therefore, proceeds by notes clearly separated from one another.

A collection of notes whose vibration-numbers bear definite ratios to one another forms a **musical scale.**

The notes of a musical scale are selected according to this fundamental law. **Simultaneous or successive notes are agreeable to the ear only when their vibration-numbers bear simple ratios to one another.**

When the ratios are not simple, audible beats occur
and dissonance results.

4. Harmonic Scale.

The harmonic scale is composed of a series of notes
whose vibration-numbers are proportional to the natural
numbers

$$1, 2, 3, 4, 5 \ . \ . \ . \ . \ .$$

The first six at least of these form agreeable harmonies
with the first.

The first is called the **fundamental** note, and the
others when heard as auxiliaries to it are called **harmonics.**

The ratio between the vibration-number of one note
and that of its antecedent note is called a **musical interval. The interval between two notes, therefore, is
obtained by dividing the vibration-number of the one
note by that of the other.**

Since the intervals between the notes are very great,
music in which the notes used form a harmonic series
is restricted and monotonous.

Following the principle that the less complicated the
ratios of the vibration-numbers of the notes the more
perfect the harmonies, physicists have constructed a
natural scale in which the intervals are much smaller.

5. Diatonic Scale.

The interval between two notes whose vibration-numbers are in the ratio $1 : 2$ is called an **octave.**

The sound produced by the simultaneous production of more than two separate notes is called a **chord.**

It is found that any three notes X, Y, Z, whose vibration-numbers are p, q, r respectively, are concordant if

$$p : q : r :: 4 : 5 : 6.$$

Three such notes are called a **harmonic triad,** and if sounded with a fourth, which is the octave of the first, they form what is called a **major chord,** the most consonant chord found in music, the ratios of the vibration-numbers being the simplest possible.

The letters C, D, E, F, G, A, B, are used to denote notes connected in harmonic triads as follows:

$$C : E : G :: 4 : 5 : 6$$
$$G : B : 2 D :: 4 : 5 : 6$$
$$F : A : 2 C :: 4 : 5 : 6.$$

Therefore,

The vibration-number of E $= \frac{5}{4}$ that of C.

"   "   G $= \frac{6}{4}$ or $\frac{3}{2}$ that of C.

"   "   B $= \frac{5}{4}$ that of G $= \frac{5}{4} \times \frac{3}{2}$ or $\frac{15}{8}$ that of C.

"   "   D $= \frac{3}{4}$ that of G $= \frac{3}{4} \times \frac{3}{2}$ or $\frac{9}{8}$ that of C.

"   "   F $= \frac{4}{3}$ that of C.

"   "   A $= \frac{5}{3}$ that of C.

Or,

the notes,   C, D, E, F, G, A, B, C, have

the vibration-ratios 1, $\frac{9}{8}$, $\frac{5}{4}$, $\frac{4}{3}$, $\frac{3}{2}$, $\frac{5}{3}$, $\frac{15}{8}$, 2 and

the intervals $\frac{9}{8}$, $\frac{10}{9}$, $\frac{16}{15}$, $\frac{9}{8}$, $\frac{10}{9}$, $\frac{9}{8}$, $\frac{16}{15}$.

The above scale is called the **natural** or **diatonic** scale.

6. Intervals of the Diatonic Scale.

The intervals in this scale are not equal. Three are represented by $\frac{9}{8}$, two by $\frac{10}{9}$, and two by $\frac{16}{15}$. The intervals $\frac{9}{8}$, $\frac{10}{9}$ are known as **tones**, the first being a **major-tone** and the second a **minor-tone**. The interval $\frac{16}{15}$ is called a **major semi-tone**.

7. Designation of Octaves.

The letters C, D, E, etc., distinguish the notes of an octave from one another. It is also necessary to have a means of designating the different octaves of any musical instrument. This is done in various ways. One of the best is to write the letter designating the note with a subscript figure which indicates the octave. For example, the C's of the eight octaves of the organ are written thus :—

$$C_{-2}, \; C_{-1}, \; C_1, \; C_2, \; C_3, \; C_4, \; C_5, \; C_6, \; C_7.$$

8. Standard of Pitch.

When once the vibration-number of any note is fixed, the vibration-ratios given above may be used to determine the vibration-numbers of the other notes. The pitch usually adopted by writers on acoustics and by makers of acoustical apparatus is $C_3 = 256$ double, or 512 single vibrations per second.

The vibration-number of the C's of the different octaves will then be

$$C_{-2}, \; C_{-1}, \; C_1, \; C_2, \; C_3, \; C_4, \; C_5, \; C_6, \; C_7$$
$$16 \quad 32 \quad 64 \quad 128 \quad 256 \quad 512 \quad 1024 \quad 2048 \quad 4096.$$

This pitch has the advantage of simplicity, the vibration-number of each C being a power of 2. The standard is a tuning-fork made to vibrate 256 times per second.

The international concert pitch adopted by musicians is $A_3 = 435$ double, or 870 single vibrations per second, and the standard is a tuning-fork made to vibrate 435 times per second.

1. Determine the vibration-number of each note in an octave when the vibration-number of A is 435.

2. What is the measure of the interval between the acoustic and the musical standard of pitch?

9. Scale of Equal Temperament.

On account of the inequality of the intervals and the consequent difficulty in constructing instruments with fixed tones, like pianos and organs, to play in different keys, the accurate diatonic scale is seldom used. A scale, called the scale of equal temperament, has been commonly adopted.

The octave is divided into twelve equal intervals, each of which is called a semi-tone, two intervals forming a tone. There are, therefore, twelve notes, and the pitch of each is obtained from the next lower by multiplying it by $\sqrt[12]{2}$; that is, the vibration-ratios of the notes in the octave are

$$1, \ 2^{\frac{1}{12}}, \ 2^{\frac{2}{12}}, \ 2^{\frac{3}{12}} \ . \ . \ . \ . \ 2$$

and each interval is $2^{\frac{1}{12}}$.

CHAPTER XVII. ·

1. The Sonometer.

The instrument called a sonometer is used in investigating the laws of vibrations of strings. Fig. 92 shows the construction of the instrument. It consists of a long, narrow resonance box, provided at its ends with steel

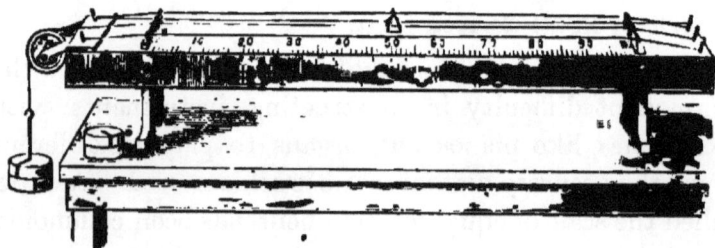

Fig. 92.

pins to which wires or strings can be fastened. The wires pass over two fixed bridges, one being placed at each end near the pins. The distance between these bridges is generally one metre. The wires are stretched either by turning the pins with a key, or by a weight. When a weight is used, the wire passes over a pulley placed at one end of the sonometer.

Movable bridges are provided to be used in changing the length of the part of any wire to be put in vibration.

[176]

2. Laws of Vibration.

Experiment 1.

Stretch two piano wires A and B on a sonometer, tune them in unison, and place a movable bridge under the centre of B. Pluck wire A at its centre and B at the centre of one of its halves. Compare the notes. It will be found that the note given by the short wire is just one octave above that given by the wire vibrating as a whole.

1. How does the vibration-number of a note given by a wire compare with the vibration-number of the note given by the same wire when its length is decreased by one-half?

2. What then is the relation between the length of a wire and the number of vibrations which it makes per second?

Experiment 2.

Place a wire B on a sonometer, let it pass over the pulley and hang a weight from the end of it. Tune another wire A in unison with it. Note the weight that is hung from B and add other weights until it vibrates in unison with (1) one-half of A, (2) one-third of A, (3) one-fourth of A, etc. It will be found that the weight is in (1) 4 times, in (2) 9 times, in (3) 16 times the original weight.

1. How does the number of vibrations per second made by the wire B when it vibrates in unison with (1) $\frac{1}{2}$ A, (2) $\frac{1}{3}$ A, (3) $\frac{1}{4}$ A compare with the number of vibrations made by the same wire when it vibrates in unison with A?

2. What caused the difference?

3. What then is the relation between the tension of the wire and the number of vibrations which it makes per second?

Experiment 3.

Place on the sonometer successively several wires B, C, D, E, etc., made of the same material, the diameters being pro-

12

portional to 1, $\frac{1}{2}$, $\frac{1}{3}$, $\frac{1}{4}$, etc. Hang the same weight from the
end of each wire, and tune a wire A to vibrate in unison with
B. Place a movable bridge under A, and slide it backward or
forward until parts of A are found which vibrate in unison
with C, D, E, etc. It will be found that C vibrates in unison
with one-half A, D with one-third A, E with one-fourth A, etc.

What, therefore, must be the relation between the diameter of a
wire and the number of vibrations which it makes per second?

Experiment 4.

Stretch with equal tension on the sonometer a steel wire and
a brass one of the same diameter. Place a movable bridge
under each, adjust the bridges until lengths of the wires
are found which vibrate in unison, and measure these lengths.
It will be found that the length of the steel wire is to the
length of the brass wire as the square root of the density of
the steel is the square root of the density of the brass; but
the vibration-numbers of wires are inversely proportional
to their lengths, therefore their vibration-numbers are inversely proportional to the square roots of their densities.

The same is found to be true of wires of other materials.

The above experiments, and others of a more general
character, carefully performed, verify the following laws:

3. Laws of Transverse Vibrations of Strings or Wires.

1. When the tension is constant the number of vibrations
per second varies inversely as the length.

2. The number of vibrations per second varies as the square
root of the tension.

3. The number of vibrations per second varies inversely as
the diameter.

4. The number of vibrations per second varies inversely as
the square root of the density of the material of which the
string is composed.

QUESTIONS.

1. What are the proportionate lengths of a stretched string which gives, when vibrating, a harmonic series of notes?

2. Stretch a wire on the sonometer, and, taking the note which it gives when it vibrates as a whole as a fundamental note, shift the movable bridge to the proper position to produce in order the other notes of a harmonic series.

3. What are the proportionate lengths of a stretched string which gives the notes of the diatonic scale?

4. Stretch a wire on a sonometer and tune it to vibrate as a whole in unison with a C-fork. Determine the positions where the movable bridge must be placed to give each of the other notes in the octave. Place the bridge in the proper positions and produce the notes.

5. A wire stretched on a sonometer by hanging a weight W from one of its ends gives the note C. What weights must be hung in succession from its end that the string may give in order the notes of the diatonic scale?

6. A string stretched on a sonometer gives a certain note. What must be the diameters of wires of the same material and the same length that will, when stretched to the same tension, give notes which will be in a harmonic series with the first?

7. A and B are two wires of the same material and thickness. A is two feet long, and is stretched by a weight of $8\frac{1}{2}$ pounds. B is four feet long, and is stretched by a weight of 34 pounds. How are the notes which the wires yield when struck related to each other?

8. A steel wire one yard long, and stretched by a weight of 5 pounds, vibrates 100 times per second when plucked. What must be the tension of two yards of the same wire that it may vibrate twice as fast?

9. Two precisely similar strings A and B have the same tension. If the tension of A is doubled, and the length of B is halved, how must the tension of B be altered to give the same note as A?

10. Two similar wires of the same length are stretched, the one by a weight of 4 pounds and the other by a weight of 9 pounds. What is the interval between the notes which they produce?

11. A stretched string 3 feet long gives the note C when vibrating transversely. What note will be given by a string one-quarter the thickness and one foot long, made of the same material and stretched by the same weight?

12. Four exactly similar strings, stretched with the same tension, are vibrating side by side. How will the note emitted be affected if they are fastened together so as to form one string by winding around them an extremely thin piece of silk?

13. A silver and an iron wire of the same diameter are stretched by weights of 4 and 36 pounds respectively. When plucked they give the same note. If the density of silver is 10.5, and that of iron 7.8, find the relative lengths of the wires.

4. Nodes and Loops.

Experiment 5.

Stretch a string on a sonometer and damp it at the centre by touching it lightly with a feather. Place a rider, made by folding a piece of paper into the form shown in Fig. 93, at the centre of one of the halves, and bow the string at the centre of the other half (Fig. 94).

Fig. 93.

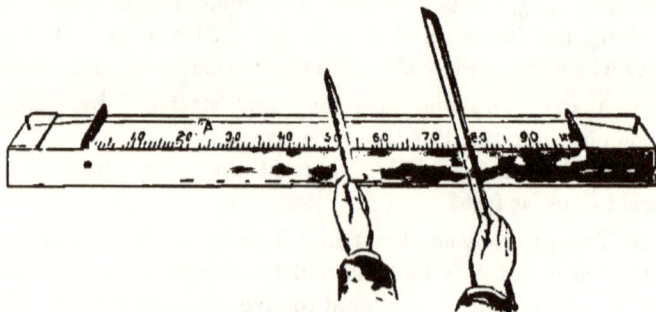
Fig. 94.

1. How does the rider behave?
2. How is the string vibrating?

Experiment 6.

Repeat the above experiment, damping the string at one-third its length from one end, placing riders on the string in the positions shown in Fig. 95, the middle one being at a point one-third of the length of the string from the end. Bow the string at the point shown.

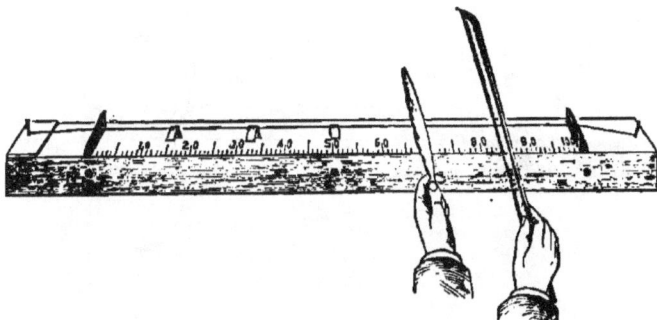

FIG. 95.

1. How do the riders behave?

2. Where are the points of least motion in the string? Where the points of greatest motion?

3. How is the string vibrating?

Experiment 7.

Repeat the last experiment, damping the string at a point (1) one-fourth, (2) one-fifth of its length from one end.

1. How does the string vibrate in each case?

2. How does the note which the string yields differ from that which it gives when it vibrates as a whole?

The above experiments show that a stretched string may vibrate not only as a whole, but may be made also to divide itself into a number of equal parts, each of which vibrates as an independent string (Fig. 96).

The points of no vibration are nodes or nodal points, and the centre of the part of the string between any two consecutive nodes, is called a loop.

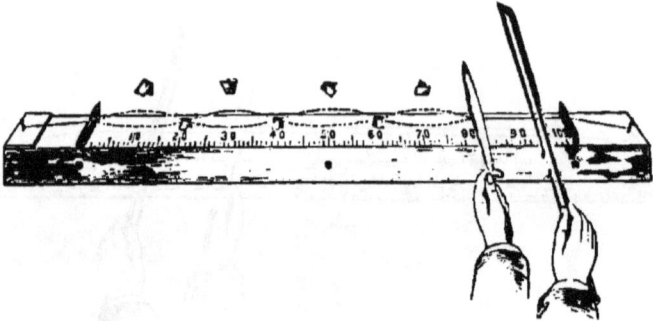

Fig. 96.

5. Overtones—Quality or Timbre.

It is impossible to vibrate a string as a whole without at the same time causing it, to a greater or a less extent, to divide and vibrate also as segments. The fundamental note of the string will, therefore, be mingled to a greater or less degree with its harmonics.

What is true of strings is also true of other sounding bodies. Smaller vibrations are superposed upon the larger, and the higher tones which they yield mingle with those given by the larger.

All the higher tones which mingle with the fundamental tones given by any sounding body are called overtones.

The combination of the overtones with the fundamental tones determines what is called the **quality**, or **timbre**, of a sound.   If the same note is produced by a flute, a piano, a violin, or by the human voice, the pitch is the same but the effect is different, although the notes may be, as nearly as possible, of the same loudness.   In fact if the same note is successively sung with the same intensity by two persons, the effect is different.   The difference is a difference in quality, and is caused by the difference in the number and the relative strength of the overtones.

The quality or timbre of a sound, therefore, is that characteristic which depends on the complexity of its vibrations.

CHAPTER XVIII.

VIBRATION OF AIR IN TUBES

I.—Resonance.

Experiment 1.

Take a glass jar about 16 inches deep of the form shown in Fig. 97. Excite a C-fork, hold it over the jar and pour water into the jar with the least possible noise.

1. What changes in the intensity of the sound take place as the water is poured into the jar?

2. What evidence have you that there is one particular length of the air-column in the tube which gives forth a maximum sound?

Repeat the experiment, using other forks.

Fig. 97.

Are the air-columns which give the maximum sounds with the different forks of equal length? If not, what is the relation between the length of the air-column and the rapidity of vibration of the fork?

The cause of the phenomena observed in the last experiment can be best understood by performing some simple experiments with the wave machine used in Exp. 2, page 143.

Experiment 2.

Fix the left-hand end of the spiral by pushing a cork into it and fastening the cork in an immovable clamp. Take hold of the right-hand end, and, by pushing it in, send a pulse of condensation along the spiral. Watch it as it is reflected from the fixed end, and the instant it reaches the free end pull the coil outward, producing a pulse of rarefaction. When this is reflected and returns to the free end, push the coil inward, and so on until the spiral vibrates as a whole steadily. You will then observe:

1. That at the fixed end the coils are alternately crowded together and then drawn apart.

2. That the amplitude of vibration of the coils at the free end is greater than at any other part of the spiral, but that they remain at about the same distance apart.

3. That to form a complete wave of condensation and rarefaction, a pulse of condensation travels from the free end to the fixed end, and back again to the free end; and is then followed by a pulse of rarefaction, which also travels from the free end to the fixed end, and is reflected back to the free end.

The wave-length is therefore four times the length of the spiral.

The motion of the coil spring furnishes an illustration of the manner in which the air-column in the tube is believed to vibrate.

The movement of a prong of the fork in the direction a to b (Fig. 98) produces in the air a pulse of condensation, which runs down to the bottom of the jar and is reflected back. Now, if the

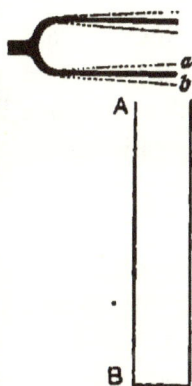

FIG. 98.

distance AB is such that this pulse of condensation

reaches the prong of the fork at the instant that it is on the point of returning from b to a, a pulse of rare-faction will be produced, which will run to the bottom of the jar and back, and, overtaking the prong when it reaches its limit a, will again be changed by it to a pulse of condensation.

The vibrations of the tuning-fork will therefore be reinforced by synchronous vibrations of the air-column in the tube, and the intensity of the sound thus increased.

It is evident that the distance AB must be just one-quarter of the wave-length of the wave produced by the fork, and since the wave-length depends upon the vibration-frequency of the fork, the distance AB must vary with different forks.

When the sound produced by a vibrating body is reinforced by the vibrations which are produced in another body, tuned to vibrate in unison with it, the effect is called resonance.

Since there can be no gain in energy, the increase in the intensity of the note when a resonator is used must be accompanied by a decrease in the length of time during which it sounds.

1. Resonators.

Resonators are used for analyzing composite sounds. Fig. 99 · shows two common forms of the instrument. They are made in various sizes, and each is carefully tuned to a definite pitch. The small opening a is placed in the ear, and if the sound to which the resonator is tuned exist in the air, it is reinforced by the vibration of the air within the cavity, and its presence is thus made known to the observer. By applying successively different resonators to the ear, the simple notes which make up a composite tone may be determined.

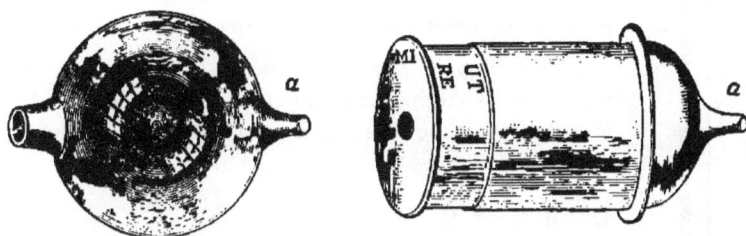

2. Determination of the Velocity of Sound by Resonance.

Since AB (Fig. 98) is just one-quarter the wave-length of the sound-wave produced by the fork, the wave-length is known when AB is measured; and if the vibration-number of the fork is known, the velocity of sound in air can be calculated from the equation

$$v = n\lambda \qquad \text{(Art. 9, page 152.)}$$

The velocity in other gases may be determined in the same way.

II.—Vibration of Air in Tubes.

3. Vibration of Air in Closed Tubes.

Experiment 1.

Take a glass tube, about 30 inches long and three quarters of an inch in diameter, insert by means of a stiff wire a cork which will slide up and down in the tube, just touching the sides (Fig. 100). Adjust the cork so that the air within the upper end of the tube will vibrate in unison with a tuning-fork. Blow across the end of the tube.

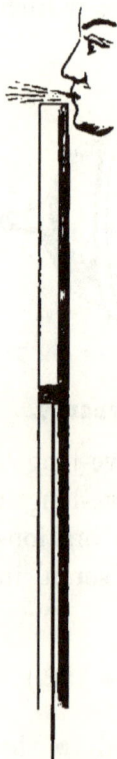

1. How does the note produced compare with that given when the fork is placed above it?

2. The blowing across the mouth of the tube causes a mixture of vibrations of different frequencies. Why is it that the tube causes one note to become predominant?

Experiment 2.

Adjust the cork in the tube used in the last experiment so that the air-column will be 24 inches in length. Blow across the end of the tube. Repeat the experiment, making the length of the air-column (1) 12 inches, (2) 6 inches.

1. What is the relation between the vibration-numbers of the notes given by (1) the 24-inch and the 12-inch air-columns, (2) the 12-inch and the 6-inch air-columns?

When a flutter, caused by the co-mingling of a number of vibrations of different frequencies, is made at the mouth of a tube, the air-column within the tube selects the vibrations which are in synchronism with itself, vibrates in unison with them, and reinforces them, thus producing a musical note.

Fig. 100.

The vibration-number of the note produced by a vibrating air-column within a tube varies inversely as the length of the tube.

4. Vibration of Air in Open Pipes.

Experiment 3.

Blow a puff of air across the end of a glass tube open at both ends. Now close one end and blow a puff of air across the open end.

How does the pitch of the note in the first case compare with that in the second case?

Experiment 4.

Select two glass tubes of the same bore, one of which is twice the length of the other, close one end of the short tube, and blow a puff of air across an open end of each.

What is the relation between the vibration-numbers of the notes produced by the two pipes?

A tube open at both ends gives a note whose vibration-number is double that given by a tube of the same length which is closed at one end.

A tube closed at one end will, therefore, give the same note as an open one of twice the length; but the wavelength of the sound-wave is four times the length of a stopped tube, **therefore the length of the open tube is one-half the wave-length of the sound-wave produced by it.**

The vibration of the air in an open tube may also be illustrated by a coil-spring wave-machine.

Experiment 5.

Push both ends inward at once, thus sending two pulses of condensation towards the centre. Watch them as they cross in the centre, and when they reach the ends, pull both ends outwards. Repeat this process until the two halves of the coil vibrate steadily in and out. Now observe:

1. That when a pulse reaches the end of the spiral its type is changed, and if it is a pulse of condensation it is reflected as a pulse of rarefaction, and *vice versa*.

2. That the centre remains stationary.

3. That the condensations and the rarefactions are not uniformly distributed along the spiral, but are greatest at the centre and least at the ends.

4. That at the ends where the coils remain always about the . same distance apart the amplitude of vibration is the greatest.

5. That the rate of vibration is twice as great as when one end of the spiral was fixed (Exp. 2, page 185). What is the reason?

The air-column in the open tube is believed to vibrate in much the same way as the coil spring was made to vibrate in the last experiment.

Fig. 101.

The movement of the prong of the fork in the direction a to b (Fig. 101) produces in the air a pulse of condensation, which runs down to the end of the jar. Its type is then changed, and it is reflected back as a pulse of rarefaction. When it reaches the upper end its type is again changed, and it is reflected as a pulse of condensation. Now if the distance AB is such that the prong of the fork is just starting to move again from a to b at the instant that this pulse of condensation starts down the tube, the vibration of the fork will be reinforced by the vibration of the air-column. This evidently can take place only when the distance AB is one-half the wave-length of the note.

5. Nodes and Loops in Pipes.

Since the layer of air at the closed end of a stopped pipe is necessarily at rest, and since rapid alternations of condensation and rarefaction take place there, the density of the air at this point, like the relative positions of the coils at the fixed end of the spiral (Exp. 2, page 185), is constantly changing.

At the open end of a tube the air has a constant density, that of the atmosphere; consequently the air particles, like the coils at the free end of the coil spring, remain at about the same distance apart. The amplitude of the vibration of the air particles at this point is therefore at a maximum.

A node in an air-column of a pipe is a section of the column where the particles of air are at rest, but where the changes of density are the greatest.

A loop is a section of the air-column where the vibrations of the particles of air have the greatest amplitudes, but where there is no change of density.

In a stopped pipe there is a loop at the open end and a node at the closed end. In an open pipe there is a loop at each end, and a node at the centre.

The existence of nodes and loops may be shown experimentally by placing a light powder in a horizontal tube (Exp. 9, page 137). When a note is sounded the powder accumulates at the points of rest. Their existence may also be shown by certain experiments with organ pipes.

6. Organ Pipes.

Fig. 102 shows the construction of a common organ pipe. Air is forced through the tube T

FIG. 102.

into the chamber C. The compressed air escapes from this chamber by a narrow slit *ed*, and, striking against the narrow bevelled edge or lip *ab*, produces a fluttering noise. The vibrations of the flutter which are in synchronism with the air-column of the tube are reinforced by it, and a musical note, the pitch of which depends upon the length of the tube, results.

7. To Show the Existence of Nodes and Loops in an Organ Pipe.

Experiment 6.

Make a small tambourine by stretching a membrane over a circular hoop. Scatter fine sand on the membrane and by means of a string lower it slowly into an organ pipe which is producing a musical note (Fig. 103). Observe the sand.

What evidence have you of the existence of loops and nodes ?

Repeat the experiment several times, varying the force of the current of air passing through the tube.

Is there a node at the centre of the tube in each case? Are there other nodes? If so, where?

8. Overtones of Pipes.

When a pipe is blown gently it yields its fundamental note. By gradually increasing the force of the current of air the air-column is made to break up into vibrating segments, · and hence to yield overtones.

Fig. 103.

The series of overtones given by a stopped pipe differs

from that given by an open one, as will be seen from a consideration of the following conditions:

(1) There must be a loop at the open end of a tube, and a node at the closed end.

(2) Nodes and loops recur alternately.

On these conditions the following will be the simplest possible divisions of air-columns in pipes.

I. In Stopped Pipes.

The open end remains always a loop and the closed end a node (Fig. 104). If there are no other nodes and loops, the pipe yields its fundamental note only.

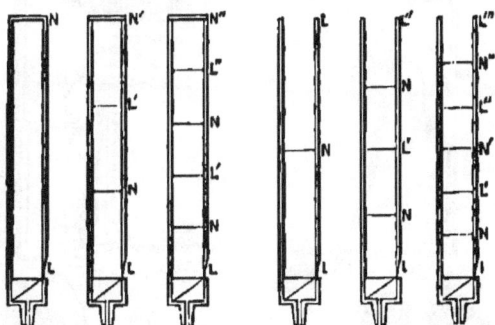

Fig. 104. Fig. 105. Fig. 106.　　Fig. 107. Fig. 108. Fig. 109.

When the pressure of the air is increased, the air-column divides into three equal parts, and an additional node and loop are formed (Fig. 105). If the air pressure is still increased, the air-column divides into 5, 7, 9, etc., equal parts (Fig. 106). Hence the vibration-numbers of the possible notes given by a stopped pipe are in the proportion 1, 3, 5, etc., and **odd harmonics only are therefore present in the overtones of a stopped pipe.**

13

II. In Open Pipes.

A loop remains always at each end of the pipe. When the fundamental note is sounded there is but one node, that at the centre (Fig. 107).

When the first overtone is produced there will be a loop at the centre, and the air-column divides as shown in Fig. 108.

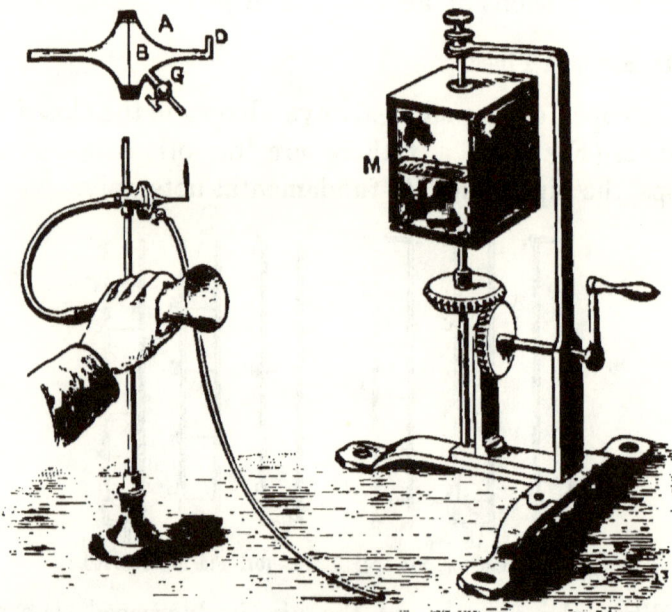

Fig. 110.

When the second overtone is produced the column will divide as shown in Fig. 109. The other overtones are formed by similar divisions. Hence the vibration-numbers of the possible notes given by an open pipe are in the proportion, 1, 2, 3, 4, etc., and, therefore, **the overtones together with the fundamental note form a complete harmonic series.**

The quality of the musical sounds given by the pipes will depend upon the degree of complexity of the vibration of the air-columns (Art. 5, page 183).

9. Manometric Flames.

Koenig has devised a means of showing the complex nature of sound vibrations by means of a vibrating gas flame. Fig. 110 shows the construction of the apparatus.

FIG. 111.

A box or capsule A is divided into two compartments by a thin membrane B. Gas is admitted into one of the compartments by a tap G, and is lit at the nozzle D. The other compartment is connected by means of a rubber

FIG. 112.

tube with a funnel-shaped mouth-piece. A rotating mirror is placed in front of the gas flame. When sound-waves enter the capsule by the mouth-piece, the membrane, gas, and flame are set in vibration. By revolving the mirror the image is drawn out into a band of light. If the flame is burning steadily, the band will be continuous; but if the flame is vibrating, it will have a wavy appearance. The complexity of the vibrations is shown by the succession of images which appear on the mirror (Fig. 111).

Fig. 112 shows the image when the vowel E is sung in front of the mouth-piece on the note C.

Experiment 7.

Revolve the mirror and sing into the mouth-piece the vowel A (1) on the note F, (2) on the note C.

Make a drawing of the image made by the flame in each case.

Fig. 113.

Let two different persons repeat in succession the above experiment.

Are the images alike? If not, explain the reason.

Vary the experiment by singing different vowel sounds on different notes, by blowing a toy trumpet and by making sounds of various kinds in front of the mouth-piece.

Instead of connecting the capsule with the mouth-piece, it may be connected with organ pipes, resonators, etc., and the character of the vibrations of air-columns observed. Fig. 113 shows a method of comparing the vibrations of two air-columns in organ pipes.

QUESTIONS.

1. Calculate the depth of a resonant jar for a fork whose vibration-number is 440, when the velocity of sound in air is 1,100 feet per sec.

2. It is found that the depth of a resonant jar which gives the loudest sound with a fork whose vibration-number is 256 is 13.2 inches. What is the velocity of sound in air?

3. A tuning-fork, making 384 vibrations per second, is held over a cylindrical jar in which the velocity of sound is 1,100 feet per second. What must be the length of the jar in order that it may be best adapted to resound to the fork? What is the length of the wave sent out by the fork?

4. When a tuning-fork is set in vibration, and held close to one end of a glass tube 20 inches long and open at both ends, an augmentation of sound takes place. If the tube is longer or shorter than 20 inches, the increase of sound is not so great. How do you explain these facts, and how could you calculate from them the pitch of the tuning-fork?

5. A stopped organ pipe 4 feet long, and an open organ pipe 12 feet long, are sounded. How are the notes related to each other? Do they differ from each other in quality, and, if so, why?

6. Give the lengths of the three shortest closed tubes, and of the three shortest open tubes, which will resound to a tuning-fork making 200 vibrations per second, the velocity of sound being 33,240 cm. per second.

7. If a pipe is constructed with holes bored in one of its sides, and these covered with little doors, as shown in Fig. 114, what effect will be produced on the vibrations of the air-column within the tube by opening A, B, and C in succession?

8. What effect is produced by the opening and the closing with the fingers of the lateral orifices of a flute? Explain.

9. A stopped pipe 2 feet long and an open pipe 4 feet long give the same fundamental notes. How do these two notes differ in quality?

Fig. 114.

CHAPTER XIX.

I.—Nature of Light.

1. Radiant Energy—Light.

The nature of radiant energy, or the energy of ether vibration, is discussed in Part I., pages 40 and 211-213.

Since by this theory, and this only, the phenomena can be satisfactorily explained, *the external physical cause of the sensation of sight is believed to be ether-waves whose wave-lengths and vibration-frequencies lie within certain limits. These waves, which have their origin in some luminous body, are propagated in all directions ; and, when transmitted to the retina of the eye, become stimuli of the optic nerve fibres, and give rise to the sensation of vision.*

Radiant energy which can affect the eye and produce vision is called Light.

As there are air-waves whose vibration-frequencies are either too slow or too rapid to affect the ear and produce the sensation of sound, so there are ether-waves whose vibration-frequencies are either too slow or too rapid to affect the eye and produce vision. These, as we shall see, are capable of producing other effects.

The vibrations of the ether take place not in the direction of the wave, like the vibrations of the particles of the air in sound-waves, but in a plane at right angles to it. In this respect they resemble water-waves (page

141), or the waves produced in the chain (Exp. 1, page 142). They are transverse waves, and not waves of condensation and rarefaction.

2. Luminous, Transparent, Translucent and Opaque Bodies.

Bodies from which light proceeds are said to be **luminous**. Those with which light originates are **self-luminous**. The sun, a lamp or gas flame, and the white-hot filament of an incandescent electric lamp are examples of self-luminous bodies.

The vibrating molecules of these bodies communicate their motion to the ether which surrounds them, and thus set up those ether-waves which have the power of exciting the optic nerves.

Bodies, like the moon, which are themselves non-luminous, become **illuminated** in the presence of self-luminous bodies by the light which they receive from those bodies and transmit by reflection.

Bodies which transmit the greater part of the light falling upon them, and through which objects can be distinctly seen, are said to be **transparent**. Those which transmit some light, but through which objects cannot be distinguished, are **translucent**. **Opaque** bodies are those which do not transmit light.

The terms transparent and opaque are relative. No body is perfectly transparent, and very thin layers of most bodies transmit more or less light.

Give examples of transparent, translucent and opaque bodies.

3. Ray, Pencil, and Beam of Light.

A **ray** is the line along which light is propagated.

A collection of rays from the same source is called a
beam, when the rays are parallel (Fig. 115), and a
pencil, when the rays are convergent (Fig. 116) or
divergent (Fig. 117). The term pencil is used by
some writers to denote any collection of rays from the
same source, and a beam is defined as a parallel pencil.

As it will be necessary, in the experiments which
follow, to use frequently a beam or a pencil of light, it will
be well at this stage to consider the means by which
these may be obtained for this purpose.

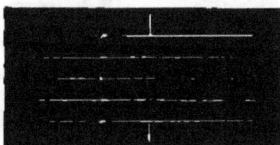

Fig. 115.

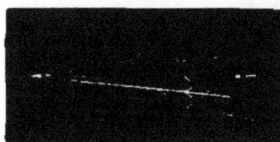

Fig. 116.

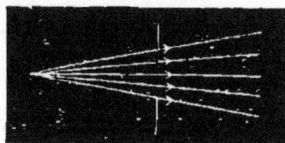

Fig. 117.

Provision must be made for darkening the windows
of the laboratory. Close wooden shutters are the best;
but blinds made of the cloth used by carriage makers
for covering buggy tops answer well. If blinds are used,
they should be mounted on spring rollers, and the light
should be shut out at the edges by having the blinds run
in grooves not less than 6 inches deep.

If the sun is the source of light, a porte lumiere is
used to transmit the light into the laboratory. This
consists of a mirror A, which can be so adjusted that
direct sunlight is reflected through the tubular opening

B (Fig. 118). A double-convex lens should be mounted in a brass ring made to slip easily into the tube B. This lens is called the condenser. A lantern objective C should be supported in front of the condenser on a metal bar D, which can be quickly adjusted or removed.

Fig. 118.

Caps E, E, with circular openings and slits are made to fit over the tube B.

A slide-holder F, which also can be attached to the tube B, as shown in the figure, should be provided.

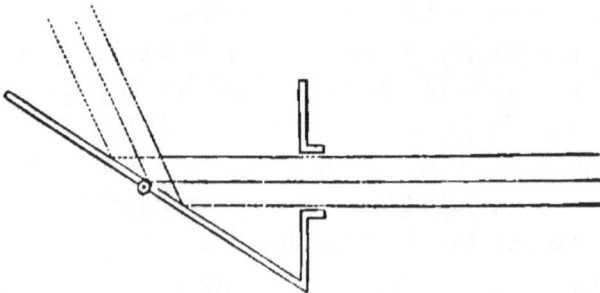

Fig. 119.

To introduce a beam of light into the laboratory, attach the porte lumiere to a shutter in a window facing south, remove all the lenses and adjust the mirror (Fig. 119). The size of the beam may be regulated by using caps with apertures of various sizes.

To introduce a convergent or a divergent pencil, slide the condensing lens into the tube (Fig. 120). The pencil will be convergent to a focus, and divergent beyond the focus.

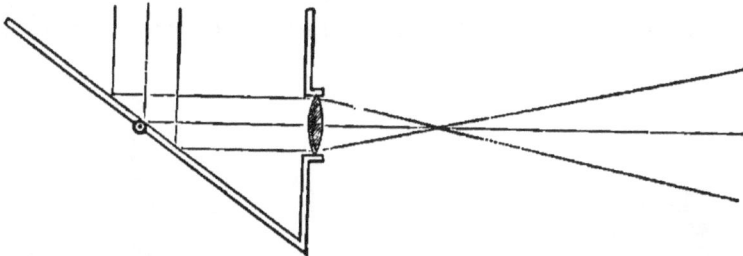

FIG. 120.

To project lantern slides, place the condensing lens and the objective in position, attach the slide-holder to the tube, place the slide in position, and move the objective backward or forward until the picture is focused on

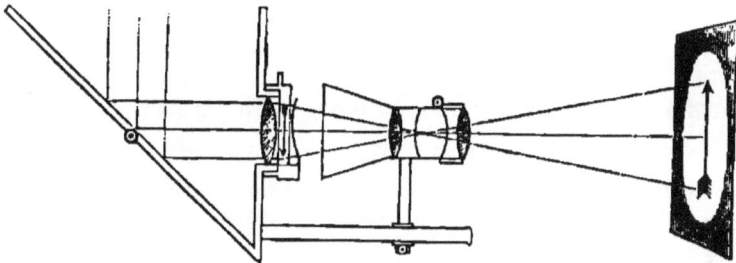

FIG. 121.

a white screen placed in front of the objective (Fig. 121). A plaster wall makes the best screen. Pieces of apparatus, and experiments that can be performed on a small

scale, may also be projected in the same manner by
placing the apparatus between the condenser and the
objective. If it is not convenient to use a lantern
objective, an ordinary double-convex lens mounted on a
stand may be used for this purpose (Fig. 5).

If the sun is not used as the source of light, a box for
shutting in the radiant must be provided. The apparatus then becomes a projection lantern. The simplest
form of the lantern is that which most nearly resembles

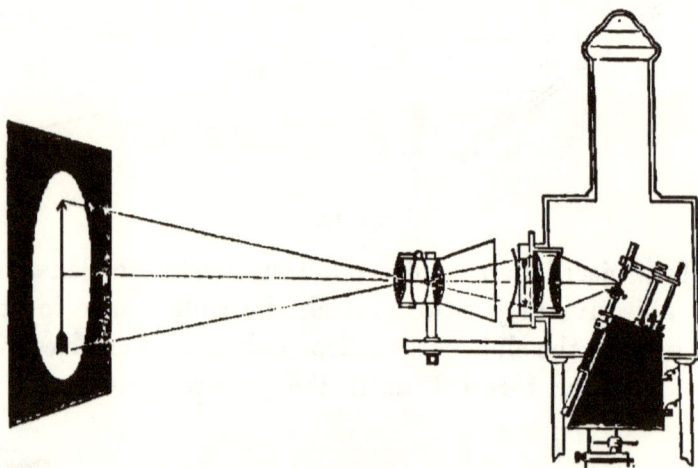

Fig. 122.

the porte lumiere described above. The tube for holding
the condensing lens, the objective, the support, the caps,
the slide-holder, etc., are the same. The lantern differs
from the porte lumiere simply in substituting a ventilated
box to enclose the radiant for the adjustable mirror which
reflects the sunlight. Fig. 122 shows a projection
lantern with an electric arc lamp as the source of
light. This is the best radiant for experimental
purposes; but an oxy-hydrogen lamp, a good coal-oil

lamp, or Auer gas burner will give sufficient light if

Fig. 123.

the room is well darkened and the screen is not placed
at too great a distance. The ordinary closed front

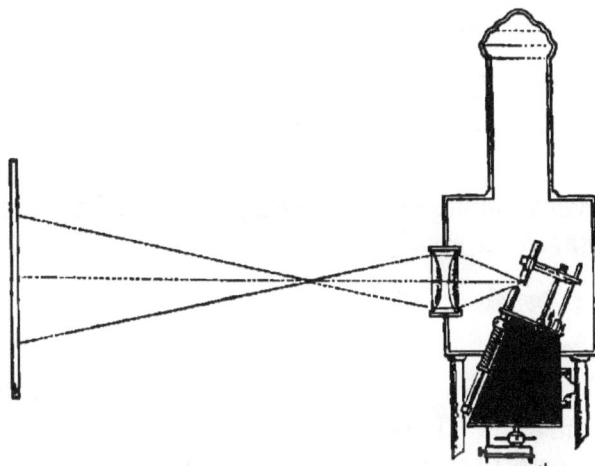

Fig. 124.

lanterns, sold for projecting slides alone, are not adapted
for physical work.

To obtain a beam of light with the lantern, remove the objective, place a single plano-convex lens in the tube, and draw the radiant back until the light becomes parallel (Fig. 123).

To obtain a convergent or a divergent pencil, place two plano-convex lenses in the tube (Fig. 124).

To project slides, place the two plano-convex lenses in the tube, rest the slide in position, and focus on the screen with the objective (Fig. 122).

II.—Rectilinear Propagation of Light.

Experiment 1.

By means of a porte lumiere or projection lantern introduce a beam of light into a darkened room, and burn some touch-paper* in its path.

1. What evidence have you that light travels in straight lines?

2. Is it the light which is made visible to you by the burning of the paper?

In every homogeneous transparent medium light is propagated in straight lines.

4. Images by Means of Small Apertures.

Experiment 2.

Make a hollow cylinder, about 6 inches in diameter, by rolling up a sheet of heavy cardboard. Place a lighted candle in a darkened room, and surround it with the cylinder. Prick a pin-hole in the cardboard on a line with the centre of the

---

* Prepared by dipping paper in a saturated solution of potassium nitrate.

flame of the candle, and place in front of the hole and at a
distance of three or four inches from it, a sheet of letter paper.

1. What is projected on the card ?

2. How is it formed, and why is it inverted ?

To answer this question consider :—

(a) That the flame is made up of an infinite number of luminous
points from each of which rays of light are passing out in every
direction.

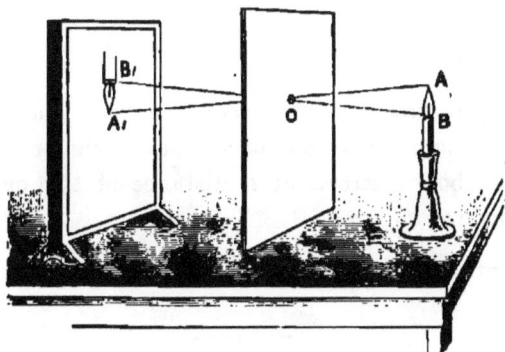

FIG. 125.

(b) That the light which passes from any one of these luminous
points A, or B (Fig. 125) through the pin-hole travels in a straight
line from the point to the hole, and continues to travel in the
same direction after passing through the hole.

(c) That when a ray of light falls upon a screen an image of the
luminous point from which it proceeds is formed on the screen.

Experiment 3.

Remove all the lenses from the porte lumiere, or projection
lantern, cover the front with a sheet of tin-foil and prick a
pin-hole in it.

A round image of the sun, or an inverted image of the
lantern radiant appears on the screen.   Make several pin-holes

near the first. Observe the number of the images increasing and overlapping as the pin-holes are made. The more the tin-foil is removed by pricking holes in it, the more the images overlap and become confused.

Remove the tin-foil altogether. The light on the screen may be regarded as the overlapping of an infinite number of images of the radiant.

Why is it that a single image is produced only by a very small aperture ?

Experiment 4.

On a bright day darken a room, make a hole about half an inch in diameter in one of the closed shutters, and place a white cardboard screen at a distance of two or three feet from the hole.

What is projected on the screen ? Explain the reason.

QUESTIONS.

1. If you make a pin-hole in the bottom of a box, and replace the lid by a piece of tissue paper, you see on the paper images of external objects. Explain the formation and the character of these images.

2. When sunlight passes through the spaces between the leaves of trees, circular patches of light are seen on the ground. Why? Are the circular patches made by openings in the leaves, which are circular? What shape would these patches have in case of a partial eclipse of the sun ?

3. Why does the size of the image (Exp. 2) become larger as the paper screen is removed further from the pin-hole ?

4. Why does the image become dimmer as it becomes larger ?

5. Shadows.

Experiment 5.

Project a large circular disc of light on a screen by placing the condensing lens in the porte lumiere or lantern. Place in the path of the light a square of blackened cardboard or tin. Observe the shadow on the screen.

Show by a diagram that the formation of the shadow is a direct result of the rectilinear propagation of light.

Move the card backward and forward.

Why does the size of the shadow projected on the screen change?

Experiment 6.

Observe carefully the outline of the shadow as formed in the last experiment, and repeat the experiment, placing the card in parallel light. Now observe the outline of the shadow.

In which case is the outline the more perfectly defined? Explain.

To answer the latter part of the last question, perform the experiment on a smaller scale as follows:—

Experiment 7.

Arrange a gas burner or coal-oil lamp to give a large, flat, fan-shaped flame. Place a white paper screen about 16 inches square at a distance of 4 or 5 feet from the flame and parallel with it. Support a card 2 or 3 inches square between the flame and the screen in such a position that a shadow about two-thirds the size of the screen is cast on it (Fig. 126). You will observe that, as in the last experiment, the part of the shadow around the edges is much lighter than that nearer the centre. To understand the reason of this, prick a pin-hole in the screen (1) in the darkest part near the centre, (2) on the line between the dark and the light parts of the shadow,

14

(3) in the light part of the shadow, (4) in the illuminated part of the screen. Place the eye behind the screen at each pin-hole, and look at the flame through each hole.

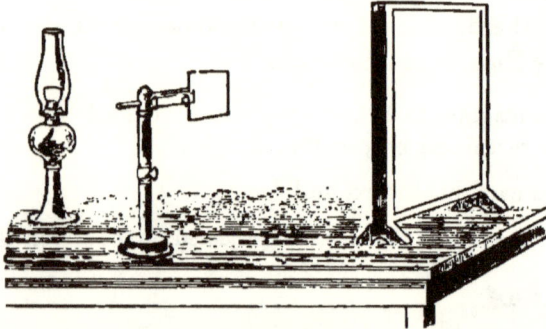

Fig. 126.

1. In which case can (1) the whole flame, (2) a part of the name, (3) the edge of the flame, (4) no part of the flame, be seen ?

2. How, therefore, do you account for the difference in the degree of darkness of the different parts of the shadow ?

The dark portion of a shadow which receives no direct light from a luminous body is called the **umbra**.

The shadow around the umbra, which is lighter than it, because it is partially illuminated by direct light from a part of the luminous body, is called the **penumbra**.

Fig. 127.

If the radiant were a luminous point there would be no penumbra (Fig. 127). The cone of light ACD, proceeding from the luminous point A, is totally stopped by the opaque body B, and the portion of the diverging cone on the side of B opposite from A will be in shadow.

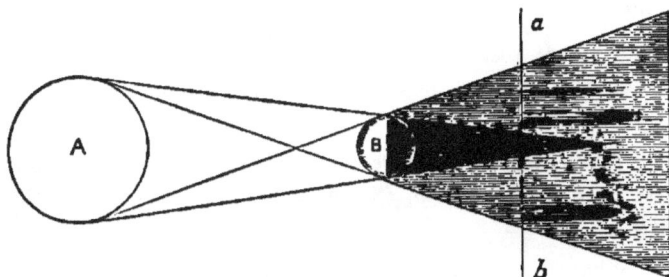

FIG. 128.

If the luminous body A is larger than the body B, the cones will be as shown in Fig. 128. A screen $a\ b$ placed

FIG. 129.

behind B will be cut by both cones, and a shadow (Fig. 129) consisting of an umbra and a penumbra will be projected on it.

QUESTIONS.

1. Make a drawing similar to Fig. 128 to show the position of the umbral and the penumbral cones when the luminous body A is (1) smaller than B, (2) equal in volume to it.

2. If a pencil is held upright between a flat gas flame and a wall, the shadow is well-defined when the flame is "edge on," but not well-defined when it is "broadside on." Explain the reason. Try the experiment.

3. Why is the shadow of a body thrown by an electric arc lamp sharp and well-defined ? Would a ground glass globe placed around the arc make a difference in the sharpness of the shadow ? Explain.

4. A person sees a partial eclipse of the sun. Is he in the umbra or the penumbra of the shadow ? Make a diagram to explain the reason for your answer.

5. When will the transverse section of an umbra of an opaque body be (1) larger than, (2) equal to, (3) smaller than, the body itself ?

6. What is the shape of a section of the earth's umbra on the moon in an eclipse ? What does this prove with regard to the shape of the earth ? Explain.

7. If a hair is held in the sunlight, about one centimetre in front of a sheet of paper, a shadow is formed ; but if it is held 5 or 6 centimetres away from the paper, the shadow becomes very indistinct or disappears. Explain.

8. A card held parallel with a wall is gradually moved from a candle towards the wall. What changes take place in the shadow ? Give the reasons for them.

CHAPTER XX.

1. Illuminating Power.

Since light is a form of energy it is a measurable quantity. It is a matter of common observation that the quantity of light given out by one luminous body may differ widely from that given out by another. A candle, for example, gives out less light than a coal-oil lamp, and a coal-oil lamp much less than an electric arc lamp. **The illuminating power of a source of light is usually measured by a unit quantity, which is the light given out by a candle of a certain weight burning at a certain rate.** The illuminating power is then measured in candle-powers.

2. Intensity of Illumination.

The **illuminating power** of a source of light must not be confounded with the **intensity of the illumination** which it produces. A candle and a coal-oil lamp although differing in illuminating power, or the quantity of light given out by them, may illuminate the same surface to the same extent, if their distances from the surface be different.

The intensity of illumination on a given surface is the quantity of light received on a unit surface.

This is manifestly dependent on :—

1. The illuminating power of the source of light, the

[213]

intensity of illumination being directly proportional to illuminating power.

2. The distance of the surface from the source of light.

Since light, like sound, travels outward in waves in every direction from its source, it is inferred by reasoning similar to that of Art. 3, page 155, that **the intensity of illumination on a given surface is inversely as the square of its distance from the source of light.**

This law may be illustrated by the following experiment :—

Experiment 1.

Remove the condensers and the objective from a projection lantern, and slip over the tube a cap in which there is a small square aperture, say 1 in. square. Place a screen successively at distances of 1 ft., 2 ft., 3 ft., etc., from the radiant, measure the side of the square of light projected on the screen at each of these distances, and calculate the areas of these squares. It will be found that they are approximately in the proportion 1, 4, 9, etc., that is, 1^2, 2^2, 3^2, etc. ; but the quantity of light falling on the screen is the same in each case, therefore the light falling on a unit surface, or the intensity of the illumination, is in the proportion 1, $\frac{1}{2^2}$, $\frac{1}{3^2}$, etc. The surface of the radiant should be as small as possible.

3. Measurement of Illuminating Power.

Since the illuminating power of any source of light is directly proportional to the intensity of the illumination which it produces on a surface when the distance is constant, and the intensity of the illumination of a surface varies inversely as the square of the distance from the source of light. the relative illuminating powers of different sources of light may be determined by comparing the intensities of the illuminations produced at known distances. All common photometers, or instruments for comparing the illuminating powers of different sources of light, depend on the application of these principles.

4. Rumford's Shadow Photometer.

Count Rumford devised a means of comparing the illuminating powers of two sources of light by a comparison of shadows. The following experiment explains his method.

Experiment 2.

To compare the illuminating powers of a coal-oil lamp and a candle, place a small upright rod in front of a screen

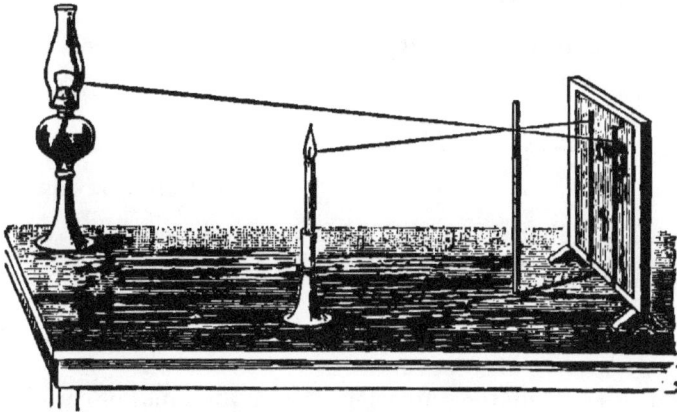

Fig 130.

made of a sheet of white paper, as shown in Fig. 130, and place the candle so that a shadow is cast by the rod on the screen when the room is darkened. Now place the lamp in such a position that another shadow of equal depth, or degree of darkness, is cast by the rod alongside the first. Measure the distance (D_1) from the candle to the screen, and the distance (D_2) from the lamp to the screen.

$$D_1 = ?$$

$$D_2 = ?$$

The part of the surface of the screen on which the candle shadow falls is illuminated by the lamp only, and the part of the surface on which the lamp shadow falls is illuminated by the candle only, and the shadows are of equal depth; hence the intensity of illumination produced by the candle when at a distance D_1 equals the intensity of the illumination produced by the lamp at a distance D_2.

If I_1 and I_2 denote the illuminating powers of the candle and the lamp respectively, I_1 and I_2 will be proportional to the intensities of the illuminations produced by the candle and by the lamp respectively at a unit distance.

When I_1 is the intensity of illumination produced by the candle at a unit distance

$\dfrac{I_1}{D_1^2}$ = the intensity of illumination at a distance of D_1

(Law of Inverse Squares),

and when I_2 is the intensity of illumination produced by the lamp at a unit distance,

$\dfrac{I_2}{D_2^2}$ = the intensity of the illumination of the lamp at a distance D_2; but the intensity of the illumination produced by candle at the distance D_1 = the intensity of illumination produced by the lamp at a distance D_2.

That is,

$$\frac{I_1}{D_1^2} = \frac{I_2}{D_2^2}$$

or

$$\frac{I_2}{I_1} = \frac{D_2^2}{D_1^2} = ?$$

If the candle is a standard candle, the illuminating power of the lamp

$$= \frac{D_2^2}{D_1^2} \text{ candle-power.}$$

5. Bunsen's Grease-Spot Photometer.

It will be observed that if a grease-spot is made on a sheet of white paper and a light placed on one side of the paper, it will appear light on a dark ground to a person on the side of the paper opposite to the light, but dark on a light ground to a person who views it from the side on which the light is situated. If the two sides are equally illuminated, the spot becomes almost invisible. The photometer introduced by Bunsen is an application of this principle.

Experiment 3.

To compare the illuminating power of a lamp flame with that of a candle by Bunsen's photometer.

Drop a little melted paraffin on a sheet of unsized paper (thin drawing paper answers well), and after it has become hard remove the excess of paraffin with a knife; place the paper between two pieces of blotting-paper and run over the blotting-paper a moderately hot iron, thus making a grease-spot on the paper about 2 or 3 centimetres in diameter. Cut out of the paper a circular disc 10 or 12 centimetres in diameter with the spot at its centre. Mount this disc in a suitable frame attached to a support.

Now draw a chalk line on a table and lay off alongside of
it a centimetre scale.   Place the candle at one end, and the
lamp at the other; and between them place the disc with the
centre of the grease-spot at the same height as each of the
flames (Fig. 131).   Move the disc backward or forward along
the line until the position is found where the grease-spot
becomes invisible when viewed in the direction of the line.

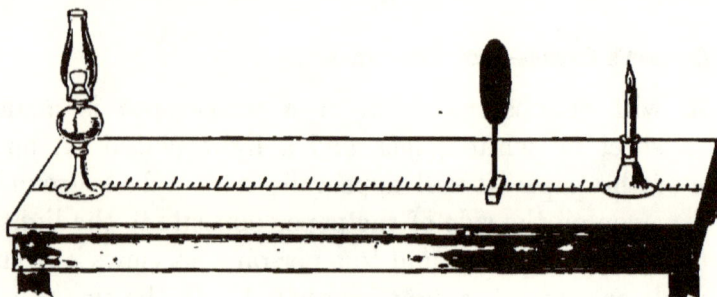

Fig. 131.

The two sides of the disc are then equally illuminated.   Measure the distance (D_1) of the candle from the disc, and the distance (D_2) of the lamp from the disc.

$$D_1 = ?$$
$$D_2 = ?$$

Let I_1 and I_2 denote the illuminating powers of the candle and
the lamp respectively.   Then, by reasoning similar to that in
Art. 4 above,

$$\frac{I_2}{I_1} = \frac{D_2^2}{D_1^2} = ?$$

In an instrument constructed for accurate determinations, the disc, candles, and supports for lights to be
tested, are mounted on holders which slide on a graduated bar.   The same bar may be made to carry lenses,

mirrors, etc., for other optical experiments. The instrument is then called an **optical bench.**

A simple form of bench may be made by fitting sliding pieces of the form shown in Fig. 131 to the trough used in Experiment 1, page 31, Part I., and attaching upright supports for holding the discs, lenses, etc., to these. A graduated scale should be placed on one side of the trough. Fig. 133 shows a bench of this form used as a Bunscn photometer.

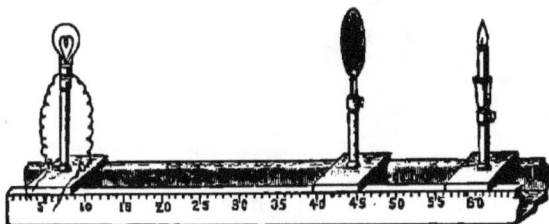

FIG. 132.

FIG. 133.

QUESTIONS.

1. A candle is placed at a distance of 2 feet from a screen, and then removed to a distance of 3 ft. Compare the intensities of illumination of the screen in the two cases.

2. A candle is placed at a distance of 10 inches from a screen and a lamp of 10 candle-power is placed on the other side of the screen at a distance of 10 feet from it. Compare the intensities of illumination on the two sides of the screen.

3. In a Rumford's photometer, it is found that the shadows are of equal depth when one of the lights is at a distance of 110 cm. from the screen, and the other at a distance of 200 cm. from it. Compare the illuminating powers of the lights.

4. How can you make use of a Bunsen photometer to prove the law of inverse squares ?

5. In measuring the illuminating powers of an incandescent lamp with a Bunsen photometer, it is found that the distance from the disc to a standard candle is 25 centimetres, and the distance from the disc to the lamp 100 cm. What is the candle-power of the lamp ?

6. A standard candle and a gas-flame of 4 candle-power are placed 6 feet apart. Where would a Bunsen disc have to be placed between them to cause the grease-spot to disappear ?

CHAPTER XXI.

I.—Laws of Reflection.

1. Experimental Verification.

We have learned (Art. 13, page 214, Part I.) that the greater part of the radiant energy falling upon bright polished bodies is reflected from their surfaces. The reflection of light may be illustrated by the following experiments.

Experiment 1.

Arrange apparatus as shown in Fig. 134. The mirror is mounted so that it can rotate on an axis. The protractor is attached to the mirror and stands at right angles to its plane, the line drawn from the axis of the mirror to the zero of the protractor being a normal to the mirror. Place the mirror so that a very small beam of light from a porte lumiere or lantern will be close to the protractor, parallel to its plane, and fall upon the mirror at its axis. Burn touch-paper in the path of the beam of light, and by turning the mirror on its axis make it take different positions.

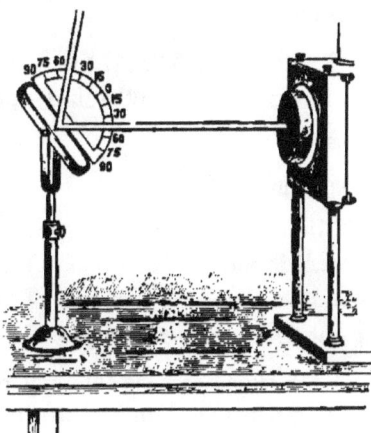

Fig. 134.

Read on the graduated

[221]

arc of the protractor the angles which the incident and the reflected beams make with the normal for each position of the mirror.

1. Are these angles always equal as the mirror is rotated ?

2. The incident beam was made parallel to the plane of the protractor, is the reflected beam also in the same plane ?

The angle which an incident ray makes with the normal to the reflecting surface is called the **angle of incidence,** and the angle which the reflected ray makes with this normal is called the **angle of reflection.**

The above experiment illustrates the following laws :—

Laws of Reflection.

1. **The angle of reflection equals the angle of incidence.**

2. **The incident and the reflected rays are both in the same plane, which is perpendicular to the reflecting surface.**

Experiment 2.

Make the laboratory quite dark, and allow a beam of light to fall in succession upon (1) a mirror, (2) a pane of polished window glass, (3) a sheet of white unglazed cardboard, (4) a sheet of blackened cardboard.

1. In which cases is the light reflected in a definite direction and a distinct patch of light thrown on a screen placed in the path of the reflected rays ? In which of these cases is the patch of light the brightest ? Why ?

2. In which of the cases is the reflecting surface seen with the greatest ease from all parts of the room ? In which is it almost invisible ? Why ?

The total amount of radiant energy falling upon a

body equals the amount reflected + the amount absorbed + the amount transmitted by the body (Art. 13, page 214, Part I.).

The mirror reflected the greater part of the light falling upon it, and transmitted and absorbed but little. The surface of the mirror being smooth, the light was reflected in a definite way and a distinct image of the radiant was projected on the screen placed in the path of the reflected rays. Since an object is seen by the light which it reflects to the eye of the observer, the mirror cannot be seen unless the eye is in the direct path of the reflected rays.

The white cardboard also reflects most of the light falling upon it; but, as its surface is rough and the various parts of it are inclined at different angles, the light is reflected in different directions, and hence scattered or diffused (Fig. 135). Since the reflected light passes outward in all directions from the

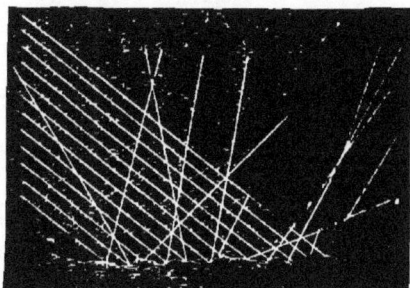

FIG. 135.

surface of this body, it can be easily seen from all points in front of its illuminated side.

The window-glass transmits the greater part of the light falling upon it; but the part reflected by it is reflected, as in the case of the mirror, in a definite way, and an image of the radiant is seen on a screen placed in the path of the reflected rays. The glass is therefore not seen from points not in the line of these rays.

The blackened cardboard absorbs nearly all of the light, and, since it reflects but little, is almost invisible.

Experiment 3.

Arrange apparatus as shown in Fig. 136, and reflect a large beam or pencil of light into a glass jar 4 or 5 inches in diameter, filled with water. Darken the room, and add a teaspoonful of milk to the water.

1. Why was the water almost invisible, while the mixture when the milk was added fills the room with radiance?

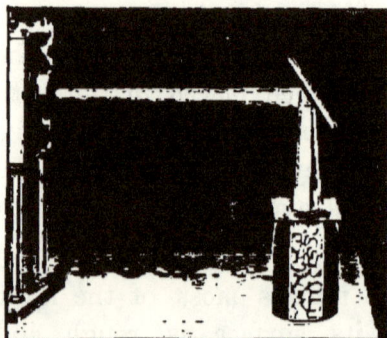

Fig. 136.

2. Why does the smoke from touch-paper make the path of a beam of light visible?

II.—Images from Plane Mirrors.

The formation of images by the reflection of light from plane mirrors is one of our most common observations.

The position of the image of a point, in relation to the mirror and the point itself, may be approximately determined by experiment in the following manner:—

Experiment 1.

Place a mirror MN (Fig. 137), which is at least 15 cm. long and 5 cm. wide, in a vertical position on a table with one of its longer edges horizontal. Stick a pin in the table at A, a point which is at a distance of about 15 cm. from the mirror. Also stick pins in the table in front of the mirror at any other points E, E_1, E_2, etc. Place the eye successively at E, E_1, E_2, and stick pins alongside the mirror at the points B, B_1, B_2, etc., B being in a straight line between E and the image of A as seen from E, and B_1, B_2 being in similar positions with respect to the image and E_1, E_2 respectively. Draw a line AC, from A at right angles to the mirror, and join by lines the pins E and B, E_1 and B_1, E_2 and B_2. Now remove the mirror and produce the lines AC, EB, E_1B_1, E_2B_2 backward until they intersect.

If the experiment is performed with care the lines will all meet in a point A_1, the distance AC being equal to A_1C.

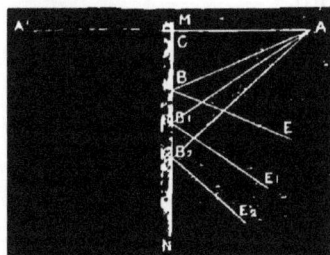

FIG. 137.

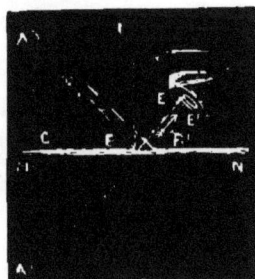

FIG. 138.

Hence

The image of a point formed by a plane mirror is behind the mirror at a distance equal to that of the point from the mirror, and on the perpendicular let fall from this point on the mirror.

3. Geometrical Construction to Find the Image of a Point Formed by a Plane Mirror.

The proposition just stated follows directly from the laws of reflection of light.

The rays of light from any luminous point A in front of a mirror MN (Fig. 138) proceed from A in all direc-

15

tions, and any ray AB incident upon the mirror is reflected by it in the direction BE, the angle of incidence ABD being equal to the angle of reflection DBE; and, to an observer whose eye is in the line BE, the light will appear to come from a point behind the mirror in the line EB produced.

If a perpendicular AC is drawn to the mirror and produced to meet EB produced in A_1, the triangle ABC will equal the triangle A_1BC, because the side CB is common to the two triangles, and the angle

$$ACB = 90° = A_1CB,$$

also the angle $$ABC = 90° - ABD$$

$$= 90° - DBE$$

$$= EBN = A_1BC,$$

therefore the triangles are equal in all respects.

Hence

$$AC = A_1C.$$

In the same way it is shown that any other ray AB_1 incident on the mirror and reflected by it along a line B_1E_1 according to the laws of reflection, will appear to proceed from the point A_1 which is in the line E_1B_1 produced.

Hence all the rays from A which fall upon the mirror appear to diverge from the point A_1. It is, therefore, the **image** of the point.

4. Virtual and Real Images.

It is manifest that the rays **only appear** to diverge from the point A_1, but the eye is affected in the same way as if the rays really did diverge from this point, and an image is seen. The image has no real existence, because the rays do not come from the other side of the mirror.

When the rays from a luminous point are so reflected, or changed, that they appear to diverge from a point, this point is called a virtual image ; but when the rays of a luminous point are so reflected, or changed, that they really diverge again from some other point, this point is called a real image.

5. Image of a Luminous Object Formed by a Plane Mirror.

The image of a luminous object is made up of the images of the infinite number of luminous points which compose it; but when the positions of the images of a limited number of these points are found, the form and the position of the image of the object can usually be determined.

For example, when the positions of the images of the points A and B of the object AB (Fig. 139) are determined by the method given in Art. 3 above, the position of the image A_1B_1 is determined.

A study of this figure shows that **the image of the object formed by a plane mirror is virtual, erect, and of the same**

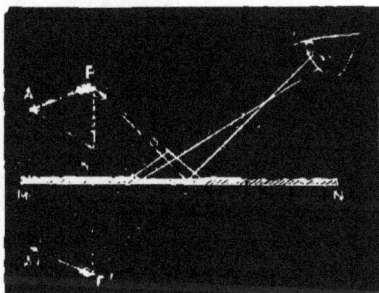

Fig. 139.

size and shape as the object, and is situated as far behind the mirror as the object is in front of it.

6. Lateral Inversion.

Experiment 2.

Stand before a mirror and hold up your right hand.

Which hand is held up by the image behind the mirror ?

Place a printed page in front of a mirror.

What peculiarity do you observe in the image of the print ? How do you account for this ?

To assist you in answering the last part of the last question, draw the image formed by a plane mirror of the letter L, by the method given in Art. 5 above, finding the positions of the images of the two extremities and the angular point.

7. Multiple Images in Inclined Mirrors.

Experiment 3.

Take two pieces of mirror glass, each 15 cm. square, and join two of their edges by pasting a strip of cloth on their backs to serve as a hinge. Set them up vertically on a sheet

FIG. 140.

of paper on which is drawn a circle graduated in degrees, the axis of the mirrors being placed in line with the centre of the circle (Fig. 140). Place a lighted candle between the mirrors, and observe the number of images formed.

1. How many images are formed when the mirrors make an angle of 90° with each other ? How many when the angle is 60° ? How many when 30° ?

2. Show by changing the positions of the mirror, and counting the number of the images formed that, if θ denotes the angle between the mirrors, and n denotes the number of images,

$$n = \frac{360}{\theta} - 1$$

when θ is an aliquot part of 360.

The formation of the images is due to the repeated reflections of the light from one mirror to the other, and each mirror gives rise to a separate series of images. The positions of the images are determined by the law that the image of a point is behind the mirror at a distance equal to that of the point from the mirror, and on the perpendicular let fall from this point on the mirror.

When the first image of the point in each mirror is determined, it is regarded as a virtual object; and its image in the other mirror is determined in the same way as if it were a real object placed at this point. This second virtual image is regarded as a virtual object, and so on. The process is repeated as long as any one of the images is situated in front of the plane in which a mirror stands.

For example, it is required to find the position of all the images of a luminous point A formed by two plane mirrors OM and ON, which make an angle of 60° with each other (Fig. 141).

Draw the line AB, at right angles to OM, and produce it to A_1, making $AB = A_1B$.

Then A_1 is the position of the first of the series of images originating with the mirror OM.

Draw a line AC, at right angles to ON, and produce it to A_2, making $AC = A_2 C$.

Then A_2 is the position of the first of the series of images originating with the mirror ON.

Now regard A_1 and A_2 as virtual objects, and find A_3, the image of A_1 in the mirror ON, by drawing $A_1 A_3$ at right angles to ON, making $A_1 B_1 = A_3 B_1$; and also find A_4, the image of A_2 in the mirror OM, by drawing $A_2 A_4$ at right angles to MO produced, making $A_2 B_2 = A_4 B_2$.

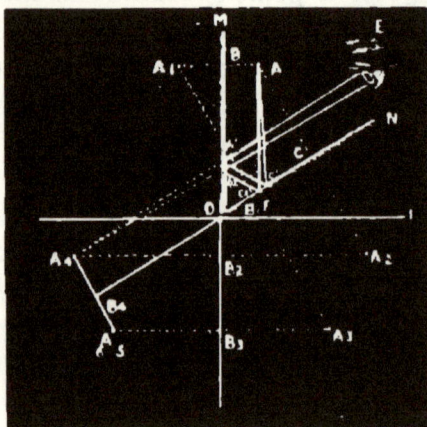

FIG. 111.

In the same way, find A_5, the image of A_3 formed by the mirror OM, and A_6, the image of A_4 formed by the mirror ON; but these images are coincident. A_5 or A_6 is behind the plane of each mirror, and no additional image is formed.

The figure also shows how to trace the rays by which any image, for example A_4, is seen by an eye placed in any position E in front of the mirrors. The light which appears to diverge from the image is in reality reflected from the surface $b_1 b_2$ of the mirror OM, upon which it is incident from the surface $c_1 c_2$ of the mirror ON, upon which, in turn, it is incident from the luminous point A. The lines representing the rays are so drawn that the angle of incidence in each case equals the angle of reflection, $b_1 c_1$ and $b_2 c_2$, if produced, intersecting in A_2.

8. Multiple Images in Parallel Mirrors.

Experiment 4.

Place two mirrors vertically on a table parallel with, and facing each other, and place a lighted candle between them. Now look obliquely into one mirror just over the edge of the other.

1. How do you account for the large number of images seen?

2. What limit is there to the number formed?

3. Why do they appear at equal distances in the same straight line? To answer this question draw the rays by which any three successive images are seen.

Experiment 5.

Hold a lighted candle before a mirror made of thick glass, and observe the images.

1. How many images are seen?

2. Explain the reason that more than one image is formed.

III.—Concave and Convex Spherical Mirrors.

9. To Determine How a Concave Mirror Disposes of Incident Light.

Experiment 1.

Cause a beam of light from a porte lumiere or lantern to be incident perpendicularly on a concave mirror, and burn some touch-paper in front of the mirror.

1. How does the mirror dispose of the light incident upon it?

The action of the mirror follows directly from the laws of reflection of light (Art. 1, page 222). But before pro-

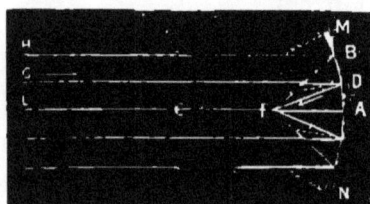

Fig. 142.

ceeding to consider this question it will be necessary to explain some of the terms applied to concave and convex mirrors. A spherical mirror MN (Fig. 142) is a very small segment of a spherical surface. It is described as **concave** or **convex**, according as the reflection takes place from the internal or the external surface. The **centre of curvature**, C, is the centre of the sphere of which the mirror is the segment.

The **centre of figure**, A, is the centre of the mirror itself.

An **axis** is any line passing through the centre of curvature and incident upon the mirror. The **principal axis** CA, is a line passing through the centre of curvature, and the centre of figure. Other axes are called **secondary**.

The **radius of curvature** is any line passing from the centre of curvature to the mirror.

To explain the phenomenon observed in Experiment 1 above, imagine the surface of the spherical mirror to be made up of an infinite number of infinitely small plane surfaces. A normal to each of these will pass through the centre of curvature.

Hence any axis of the mirror is a normal to its surface.

When a ray H, parallel with the principal axis is incident upon the mirror at B, it is reflected, and, the angle of incidence HBC being equal to the angle of reflection CBF, the reflected ray cuts the principal axis in F (Fig. 142).

When the segment MN is small as compared with the surface of the sphere, and the beam of light not large, all other rays, G, I, etc., parallel with H and the principal axis, are reflected in the same way and brought practically to the same point F. This point F is called the **principal focus** of the mirror, and the distance AF is called the **principal focal distance**.

10. Position of Principal Focus.

An incident ray HB, parallel with the principal axis, is reflected and passes through the focus F (Fig. 143).

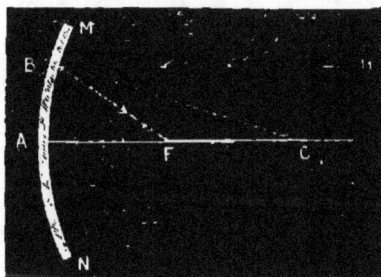

FIG. 143.

The angle   FBC = the angle HBC

                         = the angle FCB ;

therefore,   FB = FC.

When AB is very small, FB is approximately equal to FA; therefore

$$AF = FC = \tfrac{1}{2} AC, \text{ approximately.}$$

Hence

The principal focal distance for rays incident on a small portion of the surface of a spherical mirror surrounding the centre of figure equals half the radius of curvature of the mirror.

11. **To Find Experimentally the Principal Focus, and Hence the Centre of Curvature of a Concave Mirror.**

Experiment 2.

Mount the mirror on a sliding piece of an optical bench (Fig. 144), and cause a small beam of light from a lantern or

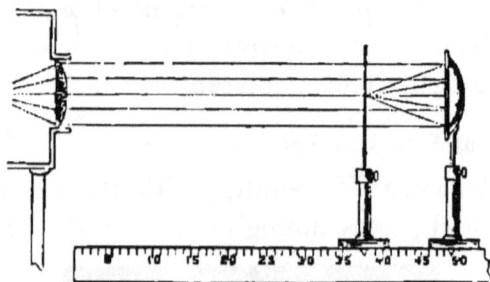

FIG. 144.

porte lumiere to be incident perpendicularly upon the mirror; attach a vertical wire to another sliding piece placed in front of the mirror, burn touch-paper in the path of the light, and move the wire up to the point where the rays are brought to a focus. Observe the distance on the scale between the wire and the centre of figure of the mirror.

1. What is the principal focal distance of the mirror?

2. What, therefore, is the radius of curvature of the mirror? Make a record of these numbers. They will be required in some of the experiments which follow.

12. To Locate Experimentally the Position of the Focus for Light Diverging from a Centre and Incident upon a Concave Mirror.

Experiment 3.

Arrange apparatus as in the last experiment. Place in the path of the beam of light near the tube of the lantern or porte lumiere a lens, the focal distance of which is about 30 or 40 cm. (Fig. 145), and attach the mirror to another sliding piece placed a distance beyond the focus of the lens. The light then which is incident on the mirror will be divergent.

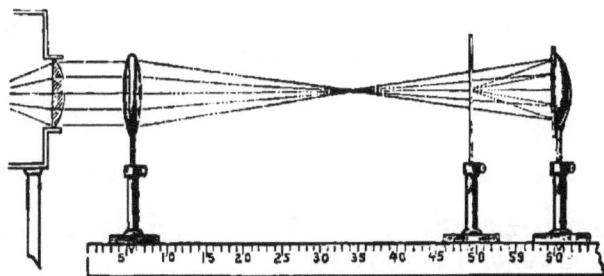

FIG. 145.

Burn touch-paper, and gradually move the mirror up toward the focus of the lens. Observe the relative positions of the focus of the lens and the focus of the mirror. They may be marked with wires as in Exp. 2.

1. Where is the focus of the reflected rays when the light diverges from (1) a point beyond the centre of curvature of the mirror, (2) the centre of curvature, (3) a point between the centre of curvature and the principal focus, (4) the principal focus, (5) a point between the principal focus and the mirror? Bring the mirror up between the lens and its focus, that is, let convergent light fall upon it. What is the course of the light after reflection?

13. Conjugate Foci.

Can the point from which light diverges and the focus to which it is brought by reflection be interchanged? Try.

Two points so related that the light diverging from either is brought by reflection from a mirror to a focus at the other, are called the conjugate foci of the mirror.

14. To Determine How a Convex Mirror Disposes of Incident Light.

Experiment 4.

Repeat Experiments 2 and 3 above, using a convex mirror instead of a concave one.

1. What changes in the direction of the rays take place by reflection when a convex mirror is placed in the path of (1) a beam of light, (2) in a divergent pencil, (3) in a convergent pencil ?

IV.—Images Formed by Concave and Convex Mirrors.

15. To Determine Experimentally the Character of the Images Formed by a Concave Mirror.

Experiment 1.

Support the concave mirror at one end of the optical bench and place a lighted candle on a sliding piece at a distance

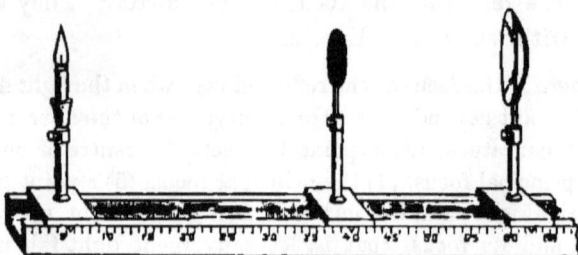

Fio. 146.

greater than the centre of curvature from the mirror. Support on another sliding piece a small paper or ground glass screen, placed between the candle and the mirror (Fig. 146). Slide the screen backward or forward until a sharply defined image of the candle is formed on it.

1 Is the image real or virtual? How do you know?

2. Is it larger or smaller than the candle?

3. Is the image erect or inverted?

4. At what point with respect to the centre of curvature and the principal focus is the image found?

Move the candle gradually toward the mirror, adjusting the screen to receive the image.

1. What change takes place in the position and the size of the image?

2. Where is the image when the candle is at the centre of curvature? Explain.

3. Where is the image when the candle is between the centre of curvature and the principal focus? What are its characteristics?

4. Where is the image when the candle is at the principal focus? Explain.

5. Where is the image when the candle is between the principal focus and the mirror? To answer this question look in the mirror. What are the characteristics of the image?

The above experiments show that in concave mirrors:—

(1) The image of an object placed beyond the centre of curvature is real, inverted, smaller than the object, and placed between the centre of curvature and the principal focus.

(2) The image of an object placed between the centre of curvature and the principal focus is real, inverted, larger than the object, and placed beyond the centre of curvature.

(3) The image of an object placed between the principal focus and the mirror is virtual, erect, larger than the object, and placed back of the mirror.

(4) The image of a luminous point placed at the centre of curvature of the mirror is coincident with the point, because the rays of light from the object return to it after reflection.

(5) No image of a luminous point placed at the focus is formed because the rays of light from it after reflection become parallel with the principal axis, and consequently are not brought again to a focus and an image formed.

How can number (4) above be used to determine experimentally the centre of curvature of a concave mirror ?

16. To Determine Experimentally the Character of the Image Formed by a Convex Mirror.

Experiment 2.

Place a lighted candle at different points in front of a convex mirror. Look in the mirror for the image. It will be seen that the image is always virtual, erect, smaller than the object, and placed back of the mirror.

17. Drawing of Images Formed by Mirrors.

In locating by a geometrical construction the points which determine the form and the position of an image, the following principles should be observed:—

1. The image of a luminous point is located where any two rays after reflection intersect.

2. The rays of which the direction after reflection can usually most easily be determined are: (*a*) a ray parallel with the principal axis, and (*b*) a ray passing in the direction of the centre of curvature. The first is reflected through the principal focus, and the second returns along the same line.

To locate the image of any luminous point, therefore, draw from the point a line parallel to the principal axis of the mirror, and from the point where this line meets the mirror draw a line through the principal focus. The image of the point will be in this line. Again, draw

from the luminous point a line through the centre of
curvature and produce it to meet the mirror. Since the
light is reflected back along this line, the image will be
in it also. Hence the image will be located at the point
where these two lines intersect.

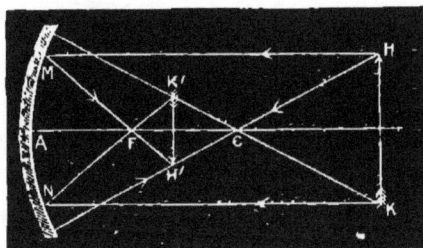

FIG. 147.

Fig. 147 shows the position and the character of the
image H_1K_1 of an object HK placed beyond the centre of
curvature of a concave mirror and Fig. 148 shows the

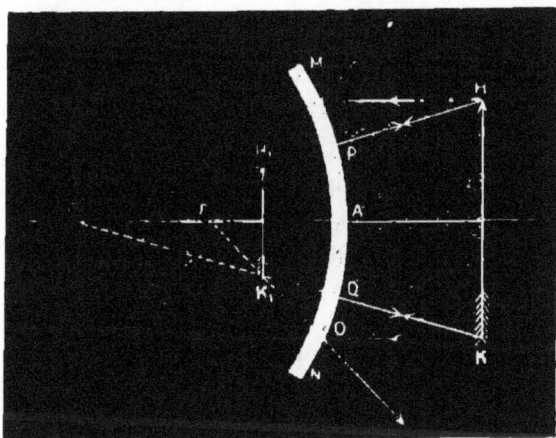

FIG. 148.

position and the character of the image of the object
placed before a convex mirror.

Make drawings showing the position and the character of the image of an object placed (1) between the centre of curvature and the principal focus, (2) between the principal focus and the mirror, (3) at the centre of curvature, (4) at the principal focus of a concave mirror.

QUESTIONS.

1. Show that when the mirror (Fig. 134) is turned on its axis, the reflected beam describes an angle which is twice as great as that described by the mirror.

2. Explain by aid of a diagram how a person can see a complete image of himself in a plane mirror one-half his height.

3. A candle stands in front of a mirror which is inclined to the vertical at an angle of 30°. Show by a diagram the position of the image, and the path of the rays by which an observer sees the two ends of the candle.

4. A person stood beside a muddy lake with the sun behind him, and his shadow was thrown distinctly on the water. He afterwards stood beside a clear deep lake with the sun likewise behind him, and saw no shadow. Explain these observations.

5. On a moonlight night when the surface of the sea is covered with small ripples, instead of an image of the moon being seen in the sea, a long band of light is observed on its surface, extending toward the point which is vertically beneath the moon. Explain this phenomenon by aid of a diagram.

6. Show by a diagram how it is possible for a lady by the use of two mirrors to see an image of the back of her head.

7. A person is equidistant from two plane mirrors which meet in the corner of a square room. In what way does the image of himself which he sees when looking toward the corner of the room differ from the image which he sees when looking toward one side of the room?

8. A ray of light is incident on one mirror in a direction parallel to a second, and after reflection at the second retraces its own course. What was the angle between the mirrors? If the ray after reflection from the second had been parallel to the first, what would have been the angle between the mirrors?

9. If you look at yourself in a convex spherical mirror you see an upright image of yourself. Under what circumstances can you see an upright image of yourself in a concave spherical mirror? Explain by means of a diagram the directions of the rays by which the images are seen.

16

CHAPTER XXII.

I.—Laws of Refraction.

1. Refraction.

Experiment 1.

Arrange apparatus as shown in Fig. 149. The front and ends of the rectangular tank are glass, and the remainder is made of metal. The sides are about 30 cm. square and 5 cm.

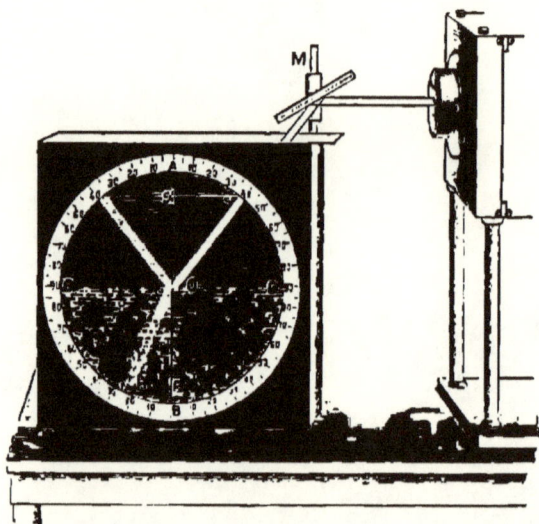

Fig. 149.

apart. The top is closed by a movable metal strip about 10 cm. longer and 4 cm. wider than the opening. A slit 3 cm. long and 1 millimetre wide is cut crosswise in this strip at a distance of about 10 cm. from the end. The whole tank.

except the glass ends and a circle about 30 cm. in diameter on the glass front, is painted inside and out a dead black. Horizontal and vertical diameters are drawn across the circular opening, and its margin is graduated in degrees, the extremities of the vertical diameter being marked zero.

The tank is filled with water to the horizontal diameter. A cap with a narrow horizontal slit is placed on the tube of the lantern or porte lumiere, and a thin beam of light is reflected by the mirror M, and made to pass through the slit in the movable top of the vessel and strike the water at O, the centre of the circle. The edge of the beam should be parallel with the front of the tank.

Observe the change in the direction of the beam at the point where it enters the water.

The change in direction which takes place in rays of light in passing obliquely from a medium of one density to a medium of a different density, for example from the air to the water in the tank, **is called refraction.**

The **incident ray** is the direction, SO, which the light takes in the first medium; and the **refracted ray** the direction, OH, which it takes in the second medium.

The angle, SOA, which the incident ray makes with a line, AB, drawn at right angles to the surface separating the two media is called the **angle of incidence;** and the angle, HOB, which the refracted ray makes with this line is called the **angle of refraction.**

The ratio, $\frac{SE}{SO}$, of the perpendicular SE to radius SO is **the sine of the angle of incidence;** and the ratio, $\frac{HF}{HO}$, of the perpendicular HF to the radius HO **is the sine of the angle of refraction.**

2. Laws of Refraction—Experimental Verification.

Experiment 2.

Repeat Experiment 1 several times, changing each time the angle of incidence by adjusting the mirror and changing the position of the slit in the movable top of the vessel. For large angles this movable strip is placed along the glass end of the vessel.

1. Observe that when the edge of the incident beam is parallel with the glass face of the vessel, the refracted beam is also in the same plane.

2. Read on the graduated scale the measure of the angle of incidence and that of the angle of refraction each time the experiment is made ; and measure in each case the length of the perpendiculars SE and HF, and calculate the value of the sine of the angle of incidence, and the sine of the angle of refraction. Tabulate your results as follows :—

No. of Experiment.	Angle of Incidence.	Sine of the Angle of Incidence.	Angle of Refraction.	Sine of the Angle of Refraction.	Ratio of the Sine of the Angle of Incidence to the Sine of the Angle of Refraction.
	In Degrees=	$\frac{SE}{SO}=$	In Degrees=	$\frac{HF}{HO}=$	$\frac{SE}{SO} \div \frac{HF}{HO} = \frac{SE}{HF} =$
1					
2					
3					
Etc.	Etc.	Etc.	Etc.	Etc.	Etc.

If the measurements are made with care, $\frac{SE}{HF}$, the ratio of the sine of the angle of incidence to the sine of the angle of refraction, will be found to be approximately equal for all angles of incidence. Carefully repeated experiments show that this law holds for other media.

Hence, from 1 and 2, we have the following laws :—

Laws of Refraction.

(1) **The incident ray and the refracted ray are in the same plane, which is perpendicular to the surface separating the two media.**

(2) **Whatever the obliquity of the incident ray, the ratio which the sine of the angle of incidence bears to the sine of the angle of refraction is constant for the same media, but varies with different media.**

3. Index of Refraction.

The ratio of the sine of the angle of incidence to the sine of the angle of refraction is called **the Index of Refraction.** It is generally denoted by the letter μ. For air and water $\mu = \frac{4}{3}$; for air and glass $\mu = \frac{3}{2}$.

4. Explanation of the Phenomena of Refraction.

The change of direction of a ray of light at the surface separating two media is a result of a change of velocity

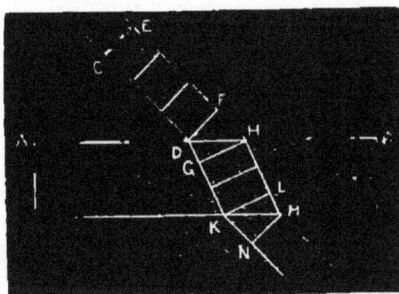

FIG. 150.

at this surface. A beam of light has a wave-front across it, that is, at right angles to the direction of its rays. Now imagine a wave whose front is DF to be incident obliquely upon a refracting surface AB, the lower medium being the denser (Fig. 150). When the ray CD has

reached the surface at D, the ray EF has reached
the point F. The ray CD is travelling more slowly in
the denser medium and, therefore, passing through a
shorter distance DG, while the ray EF is still travelling
more rapidly in the rarer medium and passing from F to
H. Hence the direction of the wave-front is changed to
the direction GH in the denser medium; and normals
to this line, GK, or HL will represent the direction of the
rays in this medium.

Again, in passing out of this medium to one of the
same density as the first, the ray GK travels more rapidly
in the rarer medium a greater distance from K to N
while the ray HL travels more slowly in the denser
medium from L to M, and consequently the direction of
the wave-front is again changed.

5. Effects Produced by Refraction.

Experiment 3.

Drop a small coin into a cup and place the eye in such a
position that it is just hidden by the edge of the cup. Keeping the relative positions of the cup and the eye unaltered,
pour water into the cup.

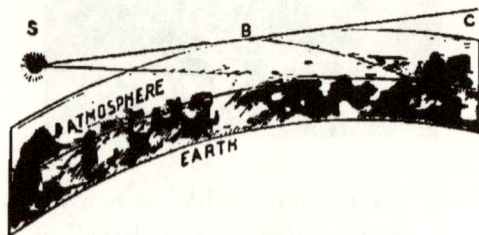

Fig. 151.

Explain why the coin is now visible, and make a drawing to
show the directions of the rays by which it is seen.

The sun is visible to us in the evening when it is in reality below the horizon; because the light which comes from it is refracted in passing into and through the atmosphere, which increases in density as it nears the surface of the earth. Fig. 151 shows the direction of the rays by which it is seen.

Experiment 4.

Plunge a straight stick obliquely into water.

1. What apparent change in the direction of the stick takes place at the point where it enters the water?

2. Make a drawing showing the directions of the rays by which the submerged part is seen.

3. Explain by means of a diagram why objects under water appear nearer the surface than they are in reality.

QUESTIONS.

1. Light falls at a given angle on a plane refracting surface for which the index of refraction is $\frac{3}{2}$. Show by a geometrical construction how to find the path of the refracted ray. (See Fig. 149.)

2. A wavy appearance is observed in the air over a hot iron. Explain the cause. Explain also the streaky appearance of water in which sugar is being dissolved.

3. If lines are drawn on paper and a piece of plate glass placed over a portion of the lines, the lines appear broken off at the edge of the glass when viewed obliquely (Fig. 152). Explain the reason.

4. You look through a thick plate of glass at a vertical pole, the glass being held so that a part of the pole is seen directly and part through the glass. Describe and explain the change in the apparent position of the part of the

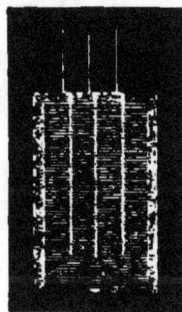

Fig. 152.

pole which is seen through the glass, when the latter is turned about a vertical axis.

5. A thick glass plate is interposed obliquely between a lighted candle and the observer's eye. Will the apparent position of the candle be altered by the glass? Explain by means of a diagram.

6. A colourless solid is dropped into a colourless liquid, and the solid is invisible in the liquid. How are the refractive indices of the liquid and the solid related? Why?

II.—Total Reflection.

Experiment 1.

Arrange apparatus as shown in Fig. 153. The movable strip is placed against the end of the tank used in Exp. 1,

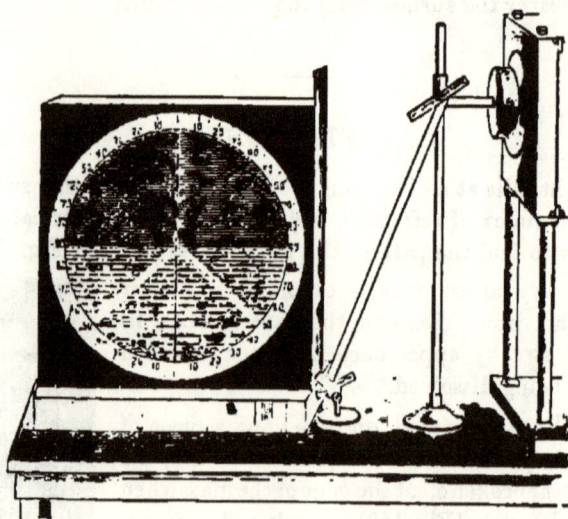

Fig. 153.

page 242. By means of two mirrors cause a thin beam of light to enter a slit placed at the **bottom** of the tank, and to pass through the water to the centre of the circle.

1. What is the course of the light after it reaches the surface of the water?

2. What is the measure of the angle which the incident beam makes with the normal?

3. The light is refracted in passing from the denser to the rarer medium; which is the greater, the angle of incidence or the angle of refraction?

Make the angle between the incident beam and the normal greater by moving the slit upward and adjusting the mirrors.

Observe that the refracted beam approaches nearer and nearer the surface of the water, and finally passes out along the surface and then into the water.

1. What angle does the incident beam make with the normal when this takes place?

When the angle of incidence is made greater than this angle the light does not leave the water, but is reflected from its upper surface as from a mirror.

6. Explanation of Phenomena of Total Reflection.

To explain the phenomena, consider Fig. 154. Since the angle of incidence is greater than the angle of refraction when a ray passes from a rarer to a denser medium, the refracted ray OH will still make an angle HOB, which is less than a right angle, with the normal, and will then be within the denser medium when the angle of

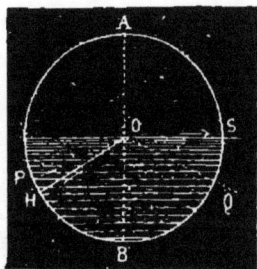

FIG. 154.

incidence is 90°, that is, when the incident ray just grazes the surface separating the two media. Now it is evident that if the process is reversed and the light is sent back from the denser to the rarer medium along the line HO,

the refracted beam will just graze the surface separating
the media, and that, if the angle of incidence is less than
the angle HOB, it will pass up into the rarer medium according to the laws of refraction. If the angle of incidence POB becomes greater than HOB the incident ray
on reaching the point O passes into the denser medium
again, or is reflected in the direction OQ, from the
surface separating the media.

The limiting angle of incidence, HOB, which allows a
ray travelling in a denser medium just to escape into a
rarer one, is called the **critical angle**.

The reflection of light from the surface of separation
of two media when the incident ray is in the denser
medium and the angle of incidence is greater than the
critical angle, is called **total reflection**.

From water to air the critical angle is 48° 35′ ; from
glass to air 41° 48′.

QUESTIONS.

1. If a beam of light is passed into a right angled glass prism,
ABC (Fig. 155), in the direction OH, it passes out in the direction

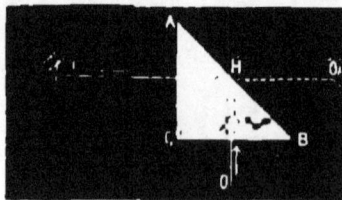

Fig. 155.

HI at right angles to OH, and the
image of any luminous object
placed at O appears at O_1. Explain. If you have a prism, place
it in the path of a beam of light
from a lantern and observe the
change in the direction of the
beam.

2. If a thick rectangular piece of glass (a paper weight answers
well) is placed on a printed page, the print can be read when the eye
is directly above it, but if the position of the eye is gradually changed,

and the page is viewed more and more obliquely, a point is reached when the print suddenly becomes invisible. Explain.

3. If an empty test-tube is thrust into water and placed in an inclined position, the immersed part appears, when viewed from above, as if filled with mercury. If the tube is now filled with water the brilliant reflection disappears. Explain the phenomena.

4. If you hold a glass of water with a spoon in it above the level of the eye and look upward at the under surface of the water, you are unable to see the part of the spoon above water, and the surface of the water appears burnished, like silver. Explain.

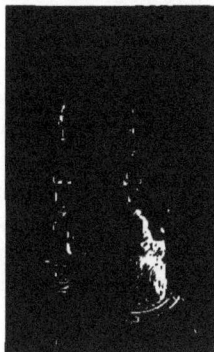

Fig. 156.

5. If a lighted candle is held obliquely before a piece of thick plate glass several images of the candle (Fig. 156) are seen. Explain, by means of a diagram, how these images are formed.

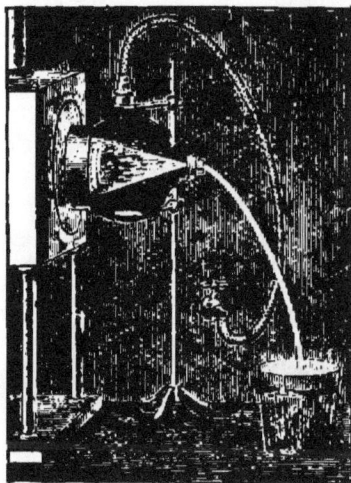

Fig. 157.

6. If a pencil of strong light is brought to a focus at the point where water is issuing in a thin stream from a vessel (Fig. 157), the

light instead of escaping into the room remains within the stream and illuminates it intensely.   Explain.

Try the experiment.   The vessel is an ordinary receiver used with retorts.   All its outer surface except the circle at which the light enters is painted black.   The water from any source of supply enters by the rubber tube, and passes out in a stream through a glass tube inserted in a cork.   The vessel is placed in such a position before the lantern or porte lumiere that the light is brought to a focus at the point where the water leaves the vessel.

III.—Refraction in Media Bounded by Plane Inclined Surfaces—Prisms.

Experiment 1.

Slide the condensers into the tube of the lantern or porte lumiere, and over the tube place a cap with a narrow vertical slit.   By means of the lantern objective or a single lens, focus the slit on the screen.

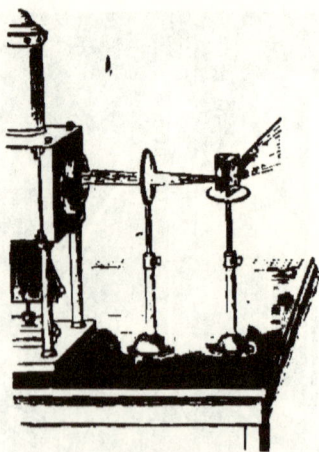

Fig. 158.

Now support a glass prism of 60° angle close to the objective lens, between it and the screen, and turn it around until the light is seen to pass through it (Fig. 158)

1. What change in direction in the light takes place in passing into and out of the prism?

2. Give a reason for this change in direction? To answer this question refer to Fig. 159, and read again Art. 4, page 245.

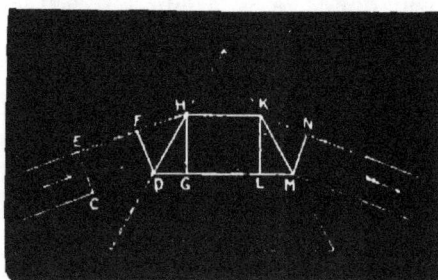

FIG. 159.

3. Why does the light emerge from this prism, while it is reflected from a side of a right angled prism, as shown in Fig. 155?

Experiment 2

Place a bright object opposite one face of the 60° prism and look at it through the prism.

How must the eye be placed to see the object? Why?

IV.—Refraction in Media Bounded by Curved Surfaces—Lenses.

A lens is any portion of a transparent medium bounded by curved surfaces. There are two classes:—

1. **Converging lenses,** those thicker at the centre than

DOUBLE-CONVEX.   PLANO-CONVEX,   CONCAVO-CONVEX.

FIG. 160.

at the edge. Fig. 160 shows the three forms of these.

All these lenses are usually called **convex.** •

2. **Diverging lenses,** those thinner at the centre than at the edge. The three forms of these are shown in Fig. 161. The term **concave** is applied to all lenses of this class.

DOUBLE-CONCAVE   PLANO-CONCAVE   CONVEXO-CONCAVE.
Fig. 161.

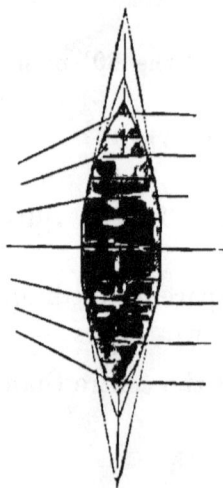

Fig. 162.

We have already observed in a number of experiments that lenses of the first class cause rays of light to become convergent. To understand the reason for this, imagine the section of the lens to be made up of a succession of prisms of gradually increasing angle placed with their bases inward (Fig. 162). The rays of light in passing through these prisms are bent towards their bases, the amount of deviation increasing with the angle of the prism. The rays, therefore, which pass through them are rendered more convergent.

In the same way, lenses of the second class may be supposed to be made up of a succession of prisms with their apices inward; consequently their effect is to render the light which passes through them more divergent.

7. Axis, Optical Centre, and Focus of a Lens.

Fig. 163 shows how a double convex lens whose surfaces are spherical is formed.

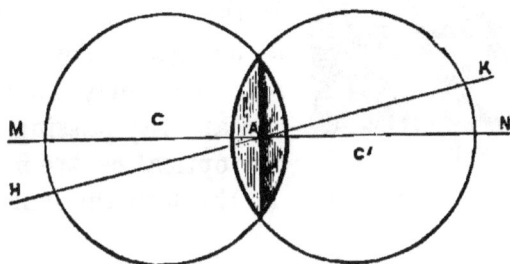

FIG. 163.

The points C, C_1 are the **centres of curvature.**

The line MN through CC_1 is the **principal axis.** The point A is the **optical centre.**

Any other line HK drawn through the optical centre is a **secondary axis.**

Construct figures to show how the other forms of lenses with spherical surfaces are formed.

The point on the principal axis of a convex lens to which rays parallel with the axis converge after passing through the lens, is called the **principal focus.** Since the rays actually converge and are brought to a focus, the focus is **real.**

With a concave lens the transmitted rays diverge and appear to come from a point in front of the lens, which is the principal focus for a lens of this class. Since these rays do not in reality come from this point, but are parallel before incidence, the focus is **virtual.**

When any ray of light passes through the optical

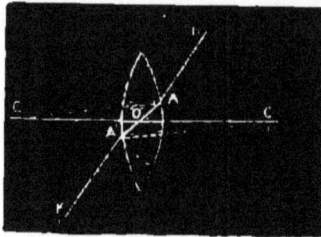

Fig. 164.

centre of a lens, the incident ray is parallel with the emergent ray, as shown in Fig. 164. If the thickness of the glass is not great the lateral displacement may be neglected, and **any ray passing through the optical centre may be regarded as passing straight through the lens without refraction.**

8. To Find Experimentally the Principal Focus of a Converging Lens.

Experiment 3.

Mount the lens on a sliding piece of an optical bench and cause a small beam of light from a lantern or porte lumiere to be incident perpendicularly upon it (Fig. 165). Mount a

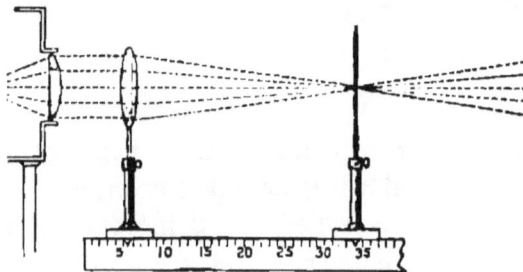

Fig. 165.

small screen on another sliding piece placed on the side of the lens opposite to the source of light. Move the screen up to the point where the spot of light falling on the screen is the smallest. Observe the distance on the scale between the screen and the centre of the lens.

What is the principal focal distance of the lens?

9. To Ascertain How a Convex Lens Disposes of a Pencil of Light.

Experiment 4.

Repeat Exp. 3, page 235, using a converging lens instead of a concave mirror.

1. Where do the rays focus when the light diverges (1) from a point beyond twice the focal distance, (2) at twice the focal distance, (3) at less than twice the focal distance, (4) at the focus, (5) from a point between the focus and the lens ?

2. What is the course of the light when the lens is placed in the path of convergent light ?

10. Conjugate Foci.

Can the point from which light diverges and the focus to which it is brought be interchanged ? Try.

Two points so related that the light diverging from either is brought by a lens to a focus at the other, are called the conjugate foci of the lens.

11. To Find the Principal Focal Distance of a Concave Lens.

Experiment 5.

Cover one of the faces of the lens with paper in which is cut a smooth round hole 2 cm. in diameter exactly over the optical centre. Mount the lens on the sliding piece of an optical bench, and cause a beam of light to be incident perpendicularly on the lens, the covered face of the lens being turned away from the light. Mount a cardboard screen on another sliding piece, placed on the side of the lens opposite to the source of light. Move the screen backward or forward until the disc of light on the screen is just 4 cm. in diameter. Observe the distance on the scale between the lens and the screen. This will be equal to the principal focal distance of the lens, as will be seen from the following considerations.

17

The rays which were parallel before reaching the lens diverge after passing through the circular opening BC (Fig. 166), and apparently come from a focus F in front of the lens. The focus is therefore virtual, and the principal focal length of the lens is AF; but when

$$B_1C_1 = 2 \ BC$$

neglecting the thickness of the lens,

$$FD = 2 \ FA$$

or　　　　　　　　　$$FA = AD.$$

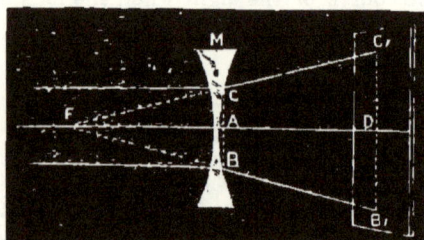

Fig. 166.

12. To Ascertain How a Concave Lens Disposes of Incident Light.

Experiment 6.

Repeat Experiments 3 and 4 above, using a concave lens instead of a convex one.

What change is produced in the direction of the rays when a concave lens is placed in the path of (1) a beam of light, (2) a divergent pencil, (3) a convergent pencil?

13. Summary.

The above experiments show that when rays from a luminous point on the principal axis of a lens fall upon it, the transmitted rays (1) converge to another point on the principal axis, or (2) are rendered parallel with the principal axis, or (3) **appear to diverge from a point on it.** When the rays transmitted actually pass through a point the focus is **real**; when they only appear to diverge from a point the focus is **virtual.**

With the convex lens the focus is real when the source of light is beyond the principal focus, and virtual when it is between the principal focus and the lens.

V.—Images Formed by Convex and Concave Lenses.

14. To Determine Experimentally the Character of Images Formed by Convex Lenses.

Experiment 1.

Support on a sliding piece of an optical bench a candle, a convex lens and a cardboard screen in the order shown in Fig. 167, their centres being in the same horizontal line.

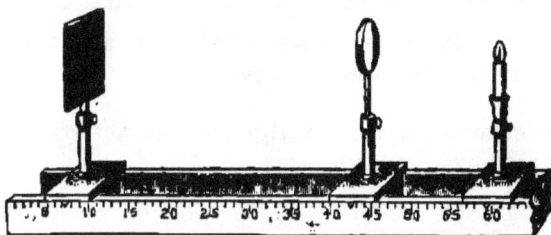

FIG. 167.

Place the candle at more than twice the focal distance from the lens. Move the screen backward, or forward, until a sharply defined image of the candle is formed on it.

1. Is the image real or virtual?

2. Is it larger or smaller than the candle? Is there any relation between the relative sizes of the candle and the image and the relative distances from the lens?

3. Is the image erect or inverted?

Move the candle gradually toward the lens, adjusting the screen as before.

1. What change takes place in the position and the size of the image?

2. Where is the image when the candle is at twice the principal focal distance? Where, when it is at the focus of the lens?

3. Where is the image when the candle is between the principal focus and the lens? To answer this question, place the eye on the side of the lens opposite to the candle and look through the lens at the candle.

The above experiments show that with convex lenses

1. The image of an object placed more than twice the principal focal distance from the lens is real, inverted and smaller than the object.

2. When the object is moved up toward the lens the image becomes larger, being equal in size to the object when it is at twice the focal distance, but remains real and inverted until the object reaches the focus, when the rays are rendered parallel and the image is at an infinite distance.

3. When the object is between the focus and the lens, the image is virtual, erect and enlarged.

15. To Determine Experimentally the Character of the Image Formed by a Concave Lens.

Experiment 2.

Look at a candle through a concave lens.

It will be found that the **image is always virtual, erect and smaller than the object.**

16. Drawing Images Formed by a Lens.

The images formed by lenses are located in a geometrical construction in the same way as the images formed by mirrors, by locating the images of certain points which determine the form and the position of the image. The

image of a point is located by finding the point of intersection after refraction of two rays proceeding from the
point. The rays of which the direction after refraction
are usually most easily determined are (a) a ray parallel
with the principal axis, which after refraction passes
through the principal focus; and (b) a ray through the
optical centre, which passes on through the lens without
change in direction. (Art. 7, page 256.)

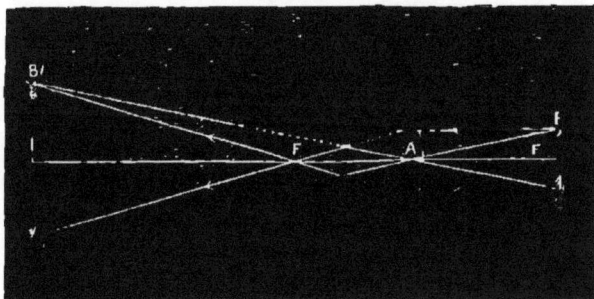

Fig. 168.

Fig. 168 shows the image formed by a convex lens
when the object is beyond the focus; and Fig. 169
shows the image formed by a concave lens.

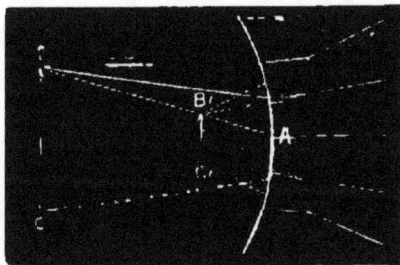

Fig. 169.

Make a similar drawing showing the image formed by a convex
lens when the object is placed between the principal focus and the
lens.

VI.—Simple Microscope.

Since the image formed by a convex lens when the object is placed between the focus and the lens is erect and enlarged, a lens of this class may be used as a simple microscope for magnifying objects placed on the side of the lens opposite to the eye. Fig. 170 shows the relative positions of the eye, the object BC, and its image B_1C_1.

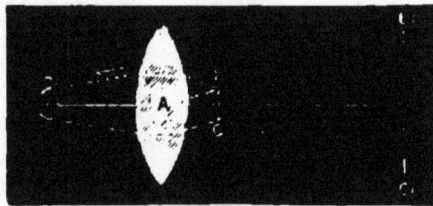

FIG. 170.

Is the image of the object seen with a simple microscope real or virtual?

CHAPTER XXIII.

DISPERSION OF LIGHT—COLOUR.

1. Decomposition of White Light—Spectrum.

Experiment 1.

Repeat Experiment 1, page 252, using a carbon disulphide bottle prism, if one is available, and placing a screen so that the rays transmitted by the prism will fall perpendicularly upon it (Fig. 171).

Observe the continuous band of colours on the screen. This is called **a spectrum.** Observe how one colour shades off into the next, passing from red at one end to violet at the other, through all the gradations of orange, yellow, green and blue.

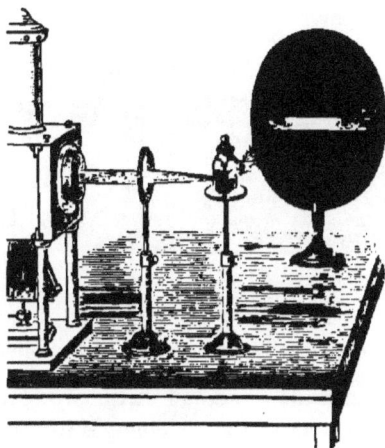

Fig. 171.

We can account for the effect observed only on the following suppositions :—

1. *That white light is composite, not simple.*

2. *That the rays of which it is composed differ in refrangibility, and consequently are separated by being transmitted through the prism.*

3. *That rays of different degrees of refrangibility give rise, when falling on the retina of the eye, to different colour sensations, the least refrangible being red and the most refrangible violet.*

[263]

Experiment 2.

Look at objects through a glass prism.

Account for the fringes of colour which appear to surround them.

2. Recomposition of White Light.

Experiment 3.

Project, as in the last experiment, a spectrum on the

FIG. 172.

screen, and place a second prism similar to the first with its apex turned in the direction of the base of the first, as shown in Fig. 172.

1. Explain why there is now projected on the screen a white image of the slit instead of the spectrum.

2. Slide a piece of cardboard along gradually between the prisms and account for the changes in the image.

Experiment 4.

Project a spectrum on the screen, and between the prism and the screen hold a large convex lens to receive the spectrum. Move it backward and forward along the line of light.

1. Is it possible to find a position of the lens where a white image of the slit is projected on the screen?

2. What does this experiment prove the coloured image to be?

Experiment 5.

Place in the slide-holder of the lantern or porte lumiere a "Newton's Disc" slide, that is a slide in which there is mounted a rotating disc (Fig. 173), in which are mounted sectors of gelatine giving the spectral colours in order. Focus the slide, and observe the colours as projected on the screen.

Now rotate the disc, at first slowly, and then gradually increase its velocity.

1. Describe what you observe.

2. How do you account for the effect produced?

FIG. 173.

These experiments show that not only can **white light be resolved into various classes of constituted rays, but also that it can be reproduced by a recomposition of these rays.**

3. Colour of Objects Due to Selective Absorption.

Experiment 6.

Again project a spectrum on the screen. Hold over the slit in succession pieces of glass of different colours, red, green, blue, etc.

Do the glasses give colour to the light, or do they quench some of the colours existing in the light? How do you know?

To more fully answer the latter part of this question, perform the following experiment.

Experiment 7.

Pass a red ribbon through the spectrum near the prism.

What colour is it in the red, in the green, and in the blue parts of the spectrum respectively?

Repeat the experiment, using (1) a white ribbon, (2) a green one, and (3) a black one.

These experiments show that **colour does not originate with the body which is said to possess it, but that it is due to the light not quenched, or absorbed, by the body.**

A body is red, because it absorbs all the rays of the white light falling upon it except those with which the sensation of red originates; it is blue, when it absorbs all the rays except those with which this sensation originates, and so on. A body is black when it absorbs all light rays.

What is a white body ?

The character of the unabsorbed rays determines the colour of a body.

The colour of a body, therefore, depends on—

1. **Its molecular structure.** Different bodies on account of difference in molecular structure absorb different classes of rays, and hence differ in colour.

2. **The nature of the light which falls upon it.** The blue ribbon placed in the red light appeared black, because the red rays were absorbed by the body which absorbs all rays except blue. For similar reasons any changes whatever in the light produce corresponding changes in the shades of colour of objects viewed by it. Bodies do not appear to have exactly the same colour in the evening as at noon-day, because the evening sunlight does not contain as much of the violet end of the spectrum as the noon sunlight, a larger proportion of the violet rays being absorbed in travelling a longer distance through the air.

3. The person who observes the body. The same rays may affect the eyes of different persons differently. While one class of rays as a usual thing affects the eye of one person in about the same way as it affects the eye of another, and hence gives rise to a similar sensation of colour; yet there are many persons who cannot make the distinctions in colour made by persons in general, for example, who cannot distinguish red from brown, orange from green, etc. Such persons are said to be colour-blind.

4. Reflection of the Rays which Determine the Colour of a Body.

When white light falls on a body the light which is reflected from its outer surface is white, because none of the rays are que·ched. The light which determines its colour enters the body, and is reflected from its internal surfaces, all other rays being absorbed by it.

5. Undulatory Theory of Light.

We are now in a position to give a more extended statement of the undulatory theory of light than that given in Chapter xix.

1. *Light is radiant energy, or the energy of ether vibration, which can affect the eye and produce vision.*

2. *Difference in colour sensation is the result of difference in the vibration-frequencies, or the corresponding wave-lengths, of the ether-waves which fall upon the retina of the eye. Ether-waves of a certain wave-length give rise to one colour, those of another length to another colour, and so on.*

3. *The vibration-frequencies of the waves which form the red end of the spectrum are less, and their corresponding wave-lengths greater,*

than those which form the violet end, the other colours being caused by waves whose wave-lengths are intermediate between these. The intermingling of waves of all these different lengths produces the sensation of white light.

4. *In a dense medium, the short waves are more retarded than the long ones, and consequently are more refracted. Hence the dispersion of light by a prism.*

5. *Ether waves are absorbed, that is, the energy of ether vibration is changed into the energy of molecular vibration, or heat, when the molecular vibrations are not in synchronism with the ether vibrations. Hence the same body will transmit one class of waves and quench another.*

6. *While only those waves whose vibration-frequencies lie between the limits of the extreme red and the extreme violet have the power of exciting the optic nerves, and producing the sensation of colour, ether-waves whose vibration-frequencies are either greater or less than these, may produce other effects. They may when falling on matter produce molecular vibrations or heat. Certain classes of them also are instrumental in bringing about chemical changes.*

To show these effects perform the following experiments.

Experiment 8.

Project a spectrum, and place in the path of the transmitted rays near the prism the face of a thermopile. Move it backward and forward across the path of the rays.

1. Does the needle of the galvanometer connected with the thermopile indicate the presence of heat on either side of the colour rays?

2. In what part of the spectrum does the thermopile indicate that the greatest quantity of radiant energy is being transformed into heat?

3. With what classes of ether vibrations are the molecular vibrations of matter most likely to be in synchronism?

Experiment 9.

Tack on a board a piece of sensitized paper, which may be obtained from any dealer in photographers' supplies. Project a spectrum and hold the board a short distance from the prism in such a position that the colours will be received on the sensitized paper. Keep it in the one position until the part of the paper on which the transmitted rays fall becomes decolorized by the chemical changes resulting from the action of the radiant energy.

1. Is there any decolorization at either end beyond the band of colours ?

2. Where is the change in colour the greatest ?

3. What classes of rays, therefore, are the most effective . in bringing about chemical changes ?

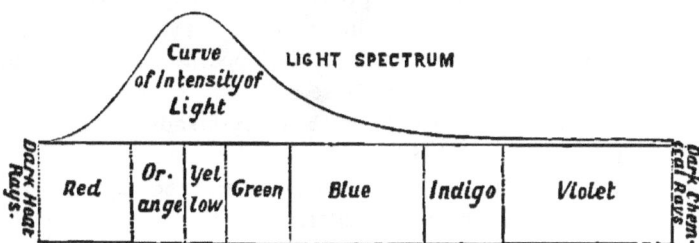

FIG. 174.

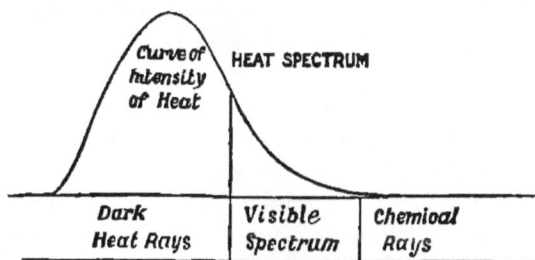

FIG. 175.

Figs. 174 and 175 show graphically the relative lighting and heating effects of the different parts of the spectrum.

It should be remembered that the truth of the undulatory theory does not rest on its power to explain simply the phenomena we have considered. By this theory, and this only, can all the phenomena connected with light and colour be satisfactorily explained.

The explanation of complementary colours, colour from interference, polarization, etc, is not within the limits of the present course.

QUESTIONS.

1. What is a spectrum? What apparatus do you require, and how would you arrange it to produce a pure spectrum from a gas flame? Give a figure showing the path of the rays.

2. Sunlight is entering a darkened room through a very narrow vertical crack in the shutter. An observer who can see the crack distinctly looks at it through a prism with its edge vertical. Describe what he sees and indicate in a figure the path of the rays to his eyes. How could he produce on the opposite wall a real image corresponding to the one which he sees?

3. Explain why a cube of glass can never show any prismatic separation of the rays.

4. A lamp flame, looked at through a glass prism, appears to be coloured blue on one side and red on the other. Draw a picture tracing the rays from the lamp to the eye, and showing which side of the coloured image is red, and which side is blue.

5. Explain the origin of colour when white light passes through a solution of copper sulphate, and trace the effect of varying the thickness traversed.

6. One piece of glass appears dark green when held up to the light, and a second piece appears dark red; explain why, when they are put together, no light passes through them.

7. A ribbon purchased by gas-light appeared to match the dress with which it was to be worn. Next morning the match appeared to be very imperfect. Explain this.

8. Light enters a room through blue glass; what appearance does a red coat present in such a room?

9. Why are sunsets characterized by red and yellow tints?

CHAPTER XXIV.

I.—Polarity.

1. Natural and Artificial Magnets.

Experiment 1.

Procure a piece of the mineral magnetite (Fe_3O_4), hammer

it out so that it is three or four times as
long as broad, plunge it into iron filings,
and lift it out (Fig. 176).

FIG. 176.

Describe what you observe.

Experiment 2.

Suspend the mineral by means of a wire stirrup and silk
fibre, as shown in Fig. 177.

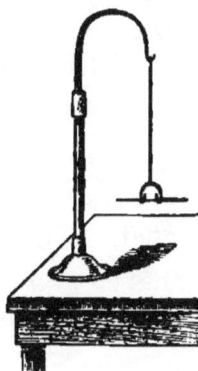

FIG. 177.   FIG. 178.

1. In what direction does it set itself? Twist it around a little
way and let it go.

2. Does it after oscillation again set itself in the same direction?

[271]

Experiment 3.

Stroke repeatedly a piece of knitting-needle from end to end in one direction with the magnetite. Now repeat Experiments 1 and 2 above, using the needle instead of the magnetite. In suspending the needle a stirrup made of very fine wire should be used (Fig. 178).

How do the results compare with those observed in Experiments 1 and 2?

A body, like the magnetite or the needle rubbed with it, which possesses the property of attracting small masses of iron to itself and of setting itself, when poised, in a definite direction pointing north and south, is called a **magnet.**

The magnetite is called a **natural magnet,** because it usually possesses magnetic properties when taken from the earth. The term **lodestone** (leading-stone) was applied to it on account of the use made of it in navigation.

A steel bar or needle which has acquired the characteristic properties of a magnet is sometimes called an **artificial magnet.** The steel is said to be **magnetized,** and the process by which it has acquired its magnetic properties is called **magnetization.**

2. The Poles of a Magnet.

It is observed in the above experiments, that the iron filings cling to a magnet in two separate tufts, one near each end, thus showing that the maximum attractive power is situated at these two points. The points are called the **poles** of the magnet, and the magnet itself is said to possess **polarity.**

The filings disappear all around the magnet midway between the poles, showing that at this line there is no attraction. This is called the **neutral line**, or the **equator** of the magnet.

An imaginary line joining the poles is called the **axis** of the magnet.

The pole which turns toward the north is called the **north-seeking** pole; and that which turns toward the south, the **south-seeking** pole.

3. Two Kinds of Magnetic Poles.

? **Experiment 4.**

Magnetize two needles, as described in Experiment 3 above, and suspend one by a fibre. Take the other magnetized needle in your hand, and hold first one end and then the other near the N-seeking pole of the suspended needle.

1. What results do you observe?

Repeat the experiment with the S-seeking pole of the magnetic needle.

2. What evidence have you that there are two kinds of magnetic poles?

4. Laws of Magnetic Attraction and Repulsion.

Experiment 5.

Suspend both the magnetic needles used in the last experiment and place them (1) so that the N-seeking pole of the one shall be near the N-seeking pole of the other; (2) that the S-seeking pole of the one shall be near the S-seeking pole of the other; (3) that the S-seeking pole of the one shall be near the N-seeking pole of the other.

What attractions or repulsions are observed?

18

This experiment verifies the following law, which is generally known as the First Law of Magnetism :—

Law I.—**Like magnetic poles repel each other, and unlike poles attract each other.**

To verify the second law with any degree of accuracy, expensive instruments are necessary. The law may be stated thus :—

Law II.—**The force exerted between two magnetic poles, whether attraction or repulsion, is directly proportional to the product of their strengths, and inversely proportional to the square of the distance between them.**

5. Magnetic Bodies and Magnets.

A **magnetic body** is any body, like iron, capable of being magnetized. A magnetic body which is not magnetized may be readily distinguished from a magnet in the following manner :—

Experiment 5.

Take one of the suspended magnetic needles, used in Experiment 4 above, and bring first to one of its poles and then to the other (1) the end of a needle that has been magnetized, (2) any part of an unmagnetized needle.

What difference do you observe in the actions of the magnetic needle in the two cases?

If the end of a body repels one of the poles of a magnetic needle it is a magnet; if it attracts both poles of a magnetic needle it is magnetic.

Several of the metals are magnetic, but only nickel and cobalt, in addition to iron, possess magnetic properties in any marked degree.

6. Neutralization of Poles.

Experiment 6.

Suspend a screw, or other small piece of iron, from one pole of a magnet (Fig. 179), and move along over this, as shown in

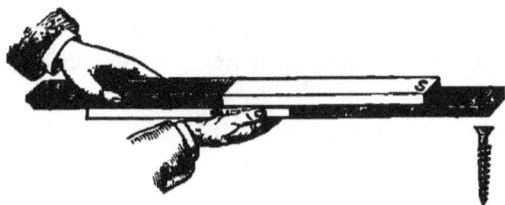

FIG. 179

the figure, another magnet of the same size and of equal power, with the opposite pole toward the screw.

What happens when the two poles are near each other?

Experiment 7.

Place the two magnets used in the last experiment end to end with opposite poles together, as shown in Fig. 180. Now

FIG. 180.

pass a suspended, or poised*, magnetic needle around the combined magnets.

What evidence have you that there is now no pole at the line where the two magnets join?

_____

* A cheap form of compass answers well for all such experiments.

Experiment 8.

Magnetize a piece of watch spring which is about 10 cm. long. Test the strength of its poles by lifting tacks with it. Now bend the spring until the two ends are in contact.

1. Does the ring thus formed show polarity at any point?

2. Two magnetic needles connected rigidly by a wire, as shown in Fig. 181, are called an **astatic pair.** In what direction will they set themselves when suspended by a fibre?

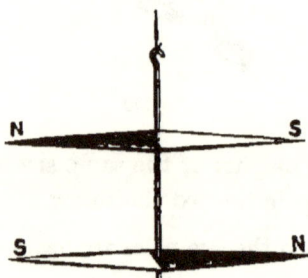

Fig. 181.

When the opposite poles of two magnets, or of the same magnet, are placed together, they tend to neutralize each other.

7. The Two Poles of a Magnet Inseparable.

Experiment 9.

Magnetize a knitting-needle, roll it in iron filings, and notice the positions of the poles. Break the needle into halves, and roll each half in the filings. Continue the process until the needle has been broken into small pieces.

1. What evidence have you that the two poles of a magnet are inseparable?

2. Are the poles of each separate piece of the needle different, that is, is one pole a N-seeking and the other a S-seeking pole?

To answer this question present each end of one of the pieces to the N-seeking or the S-seeking pole of a suspended magnetic needle.

Place together the broken ends of two halves which have been separated, and each of which consequently possesses two poles.

Do the adjacent poles neutralize each other, leaving poles only at the ends?

A magnet cannot possess one pole only. Two poles, one N-seeking and the other S-seeking, co-exist in every magnet.

8. Theories of Magnetism.

The co-existence of two poles in even the smallest part of a magnet is usually explained on the *theory that the molecules of a magnetic body possess polarity.*

When the body as a whole apparently possesses no magnetic properties, the opposite poles of adjacent molecules neutralize one another (Fig. 182a); *but when it is magnetized, the greater number of the molecules are turned into lines, with their N-seeking poles turned in one direction and their S-seeking poles in the opposite direction.*

FIG. 182a.   FIG. 182b.

When, therefore, the magnet is broken at any point, one face of the fracture is a N-seeking and the other a S-seeking pole.

If the magnetization were equal at all points of the magnet and the molecules were all arranged in line with their magnetic axes parallel to the axis of the magnet with like poles all pointing in one direction (Fig. 182b), free poles would be found only at the end surfaces; but this is not the case, because we have found that complete neutralization takes place only near the centre of the magnet. The intensity of magnetization must, consequently, be greatest at the middle of a bar of steel, and less toward its ends.

The theory of molecular polarity may be illustrated by the following experiment:—

Experiment 10.

Fill a test-tube nearly full with steel filings (Fig. 183*a*), magnetize it by drawing one pole of a strong magnet over the tube repeatedly in the same direction. Observe that the filings set themselves end-ways (Fig. 183*b*).

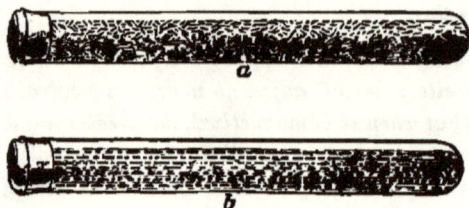

Fig. 183.

Without disturbing the arrangement, bring one end of the tube near (1) the N-seeking pole, (2) the S-seeking pole of a suspended magnetic needle.

What evidence have you that the tube filled with steel filings acts as a bar magnet?

Disturb the arrangement of the filings by shaking the tube, and again present one end of the tube to each pole of a magnetic needle.

1. Does the tube now act as a bar magnet? How do you know?

The tube filled with steel filings is shown to be a magnet when the magnetic axes of the individual pieces of steel making up the filings are parallel with the length of the tube, similar poles being turned in the same direction; but when this arrangement is disturbed, and these individual pieces of steel turn in various directions, their poles neutralize one another, and the tube as a whole no longer acts as a magnet.

In a similar manner, the magnetic action of steel is believed to depend on the arrangement of the molecular magnets of which it is built up.

Ampère's theory, which is offered as a possible explanation of the polarity of the molecules, will be stated at a later stage.

9. Consequent Poles.

All magnets must have, as we have learned, at least two poles; but, on account of irregular or imperfect magnetization, a piece of steel may be made to have one or more poles between those at the ends. These are called **consequent** poles. The magnet in this case may be regarded as consisting of two or more magnets placed end to end with similar poles together.

FIG. 184.

Experiment 10.

Take two similar magnets of equal strengths, and place them end to end with similar poles together, say two N-seeking poles together (Fig. 184). Determine the position of the poles by bringing a suspended magnetic needle in different positions (1) near each end, (2) near the point of junction.

1. What pole is found to be at each end of the joined magnets? What pole at the middle, or point of junction?

2. If two S-seeking poles were placed together, what would be your answers to the last question?

3. If three magnets were placed as shown in Fig. 185, what would be the distribution of the poles?

Fig. 185.

~ **Experiment 11.**

Place like poles of two bar magnets at the middle of a piece of watch spring and draw them simultaneously to the ends. Now lift them up, place them at the centre, and draw them to the ends. Repeat the operation several times. Remove the magnets, and determine the distribution of the poles in the watch spring by passing a suspended magnetic needle around the spring, and noting the direction in which the poles of the needle point in its various positions.

What pole is in the middle of the spring? What at each end?

QUESTIONS.

1. You are provided with a steel sewing-needle and are required to magnetize it so that its point may be a South-seeking pole. How will you do it?

2. A bar magnet, freely suspended horizontally, sets itself north and south. If a second bar magnet is suspended by the side of the first, how will they act upon each other? Make your answer clear by a diagram.

3. You are doubtful whether a steel rod is neutral, or is slightly magnetized; how could you determine its magnetic condition by trying its action upon a compass-needle?

4. A bar magnet is placed anywhere on the table in the neighbourhood of a compass needle, and is slowly rotated round a vertical axis through its middle point. Describe the behaviour of the needle (1) when the magnet is very close, (2) when it is a few feet distant.

5. Six magnetized sewing-needles are thrust through six pieces of cork, and are then made to float near together on water with their N-seeking poles upward. What will be the effect of holding (1) the S-seeking pole, (2) the N-seeking pole, of a magnet above them ? Try the experiment.

6. A bar magnet is placed anywhere on the table in the neighbourhood of a magnetic needle, and is slowly rotated round a vertical axis through its middle point. There are two positions of the magnet for which the needle points along the line of its undisturbed position. Explain this.

7. Two bar magnets of equal length are set on end a few inches apart. A small magnetic needle is carried round the upper poles in a figure-of-eight course. How will it point in the various positions occupied, (1) when the upper poles are *like* poles ; (2) when they are *unlike* poles ?

8. A strong bar magnet is placed on a table with its north pole pointing toward the north. State in what direction a compass-needle points (1) when placed immediately over the centre of the bar magnet, (2) when gradually raised vertically upwards.

9. A compass-needle is suspended at the centre of a circle drawn on a horizontal table. A magnet is moved round the compass so that its centre always lies in the circumference of the circle and its length always points east and west. How and why will the position of the compass-needle change as the magnet is carried round it ?

10. A piece of steel wire, bent so as to form two sides of a square, is magnetized in such a way that each of its free ends is a north pole, and the bend a south pole. When placed upon a cork floating in water, how will it set itself ?

11. Arrange three bar magnets of equal strength so that there is no magnetic effect on a neighbouring magnetic needle.

12. It is suspected that a magnetized bar of steel has consequent poles. How would you ascertain whether this is so or not?

13. Two equal and equally magnetized bar magnets are fastened together at the centres at right angles to each other, so as to form an equal-armed cross. How will the cross set itself when balanced on a pivot?

II.—Magnetic Induction.

10. Phenomena of Induction.

Experiment 1.

Place a piece of soft iron with one of its ends near the pole of a magnet, as shown in Fig. 186, and bring a tack, or other small piece of iron, to the other end of the soft iron.

Fig. 186.

What do you observe?

Remove the magnet.

1. What takes place?

2. What properties did the soft iron possess when near the magnet?

3. Did it permanently retain these properties?

Experiment 2.

Place the soft iron again in the same position near the magnet, and bring a suspended magnetic needle near the end of the soft iron which is the more remote from the magnet, as shown in Fig. 187.

1. What is the pole at one end of the soft iron when the other end is near (1) the N-seeking pole, (2) the S-seeking pole of the magnet?

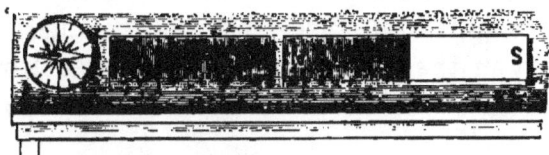

Fig. 187.

This magnetizing action of the magnet upon the soft iron placed close to it is known as induction. **Under induction, a magnetic pole induces opposite polarity in the end of a rod of soft iron nearest it, and similar polarity in the end farthest from it.**

Experiment 3.

Repeat Experiment 2 above, placing two or more small soft iron rods between the magnet and the magnetic needle.

Does induction take place through a series of iron rods?

11. Induction Precedes Attraction.

Experiment 4.

Place a strong magnet on a table with its N-seeking pole projecting over the edge. Bring a tack to this pole of the magnet, add others to this in the form of a chain, as shown in Fig. 188.

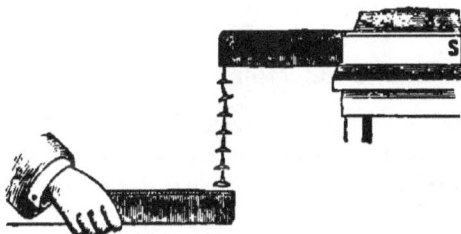

Fig. 188.

1. Explain why any one tack is held to the one above it.

2. What happens when (1) the N-seeking pole, (2) the S-seeking pole of another magnet is brought near the lowest tack ?   Why ?

A piece of unmagnetized iron is drawn to the pole of a magnet because the pole induces opposite polarity in the end of the unmagnetized iron nearest it, and these unlike poles attract each other.

< **Experiment 5.**

Place a strong magnet on a table and suspend by means of a wire stirrup and fibre a bar of soft iron over it (Fig. 189).

1. What position does it take ?   Why ?

Fig. 189.

2. What happens when (1) the N-seeking pole, (2) the S-seeking pole of another magnet is placed, as shown in the figure, near the end of the soft iron over the S-seeking pole of the first magnet? Explain.

12. Explanation of Induction.

The phenomena of induction can be explained on the theory of the polarity of the molecules. *When one of the poles, say the N-seeking pole, of a magnet is brought near the end of the rod of soft iron, the attractions and repulsions between it and the magnetic molecules of the soft iron cause these molecules to turn around and arrange themselves with their S-seeking poles turned toward the*

N-seeking pole of the magnet. The bar, like the tube containing the steel filings (Fig. 183), then shows polarity, its S-seeking pole being at the end of the rod nearest the N-seeking pole of the magnet.

When the magnet is removed, the mutual attractions and repulsions among the molecules of the soft iron cause the poles to turn again in various directions and thus to neutralize one another's action. The rod, therefore, no longer appears a magnet.

In the case of steel, which possesses greater molecular rigidity, greater difficulty is found in causing the molecules to set themselves with one class of poles pointing in one direction; but when the poles have once set themselves in this way, they retain their relative positions for a long time. For this reason, the steel remains permanently a magnet, while the soft iron possesses magnetic properties only while under the direct influence of the inducing magnet.

Steel is usually magnetized in one of the following ways:—

13. Methods of Magnetization.

1. Single Touch.

This method consists in rubbing the bar of steel to be magnetized repeatedly in the same direction with one pole of another magnet placed as shown in Fig. 190.

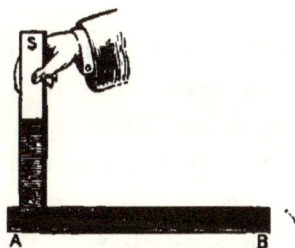

Fig. 190.

2. Double Touch.

The bar AB to be magnetized is usually supported on two magnets arranged as shown in Fig. 191, and is stroked first in one direction and then in the other by two bar magnets the opposite poles of which are kept at a constant distance from each other by means of a piece of wood, as shown in the figure.

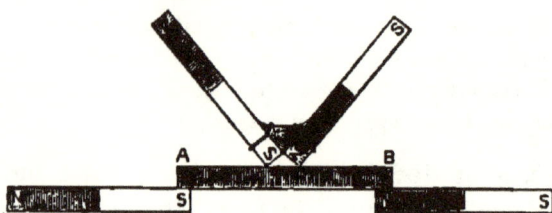

FIG. 191.

3. Separate Touch.

The bar to be magnetized is supported on the opposite poles of two magnets as in the last case. The inducing magnets are placed as shown in Fig. 192, and are drawn

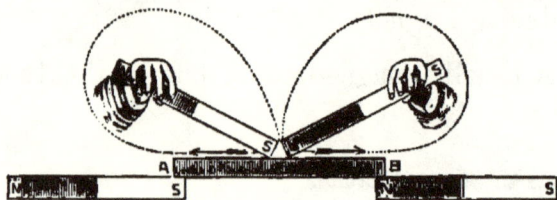

FIG. 102.

away from each other to the two ends of the bar, lifted up, carried back in a wide curve through the air, placed again at the middle, and again drawn away from each other to the two ends. This process is repeated several times.

Explain clearly by the theory of the polarity of the molecules how the steel is magnetized by each of the above methods.

4. By an Electric Current.

This will be considered at a later stage.

Anything which tends to give freedom of molecular

movement, such as hammering, twisting, heating, etc., assists in the process of magnetization by induction. Why ?

14. Retentivity.

We have observed that the retentivity, or the power to resist demagnetization, is very small in soft iron and very great in hard steel; but even in steel the mutual attractions and repulsions among the poles of the molecules tend to cause them to pair off again in such a way as to neutralize each other, and a certain amount of loss of power is inevitable if the steel is subjected to any molecular disturbance. This loss may be prevented to a

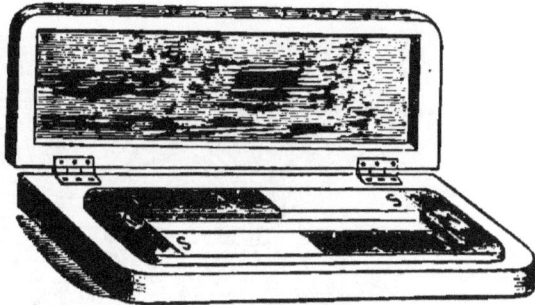

Fig. 193.   Fig. 194.

great extent by the use of "keepers." These are pieces of soft iron, which are used to join the poles of the magnet when not in use. Each pole of the magnet induces the opposite kind of polarity in the end of the keeper in contact with it, and these opposite poles attract each other and tend to preserve the arrangement of the molecules necessary for magnetization.

Fig. 193 shows a horse-shoe magnet with its keeper, and Fig. 194 indicates how keepers may be applied to bar magnets.

Experiment 6.

Magnetize a needle by single touch.   Prove that it is magnetic by rolling it in iron filings.   Heat it red hot, allow it to cool, and again test its magnetic power.

1. What do you observe ?

2. Explain the change.

15. Magnetic Field.

The space surrounding a magnet pervaded by the magnetic forces is called the field of the magnet. At every point in the field the magnetic force has a definite strength, depending, as we have seen, on the distance of the point from the poles.

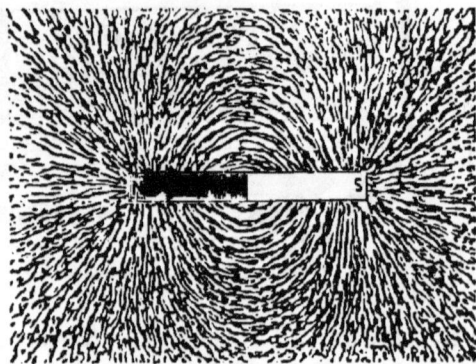

the con. lion
? iw ... r ...
r.. i.l

Fig. 19.

16. Magnetic Lines of Force.

Experiment 7.

Lay a sheet of heavy paper or cardboard on a short bar magnet, and sprinkle iron filings over the paper by sifting them through a piece of muslin.   Gently tap the paper and observe the manner in which the filings arrange themselves under magnetic induction (Fig. 195).

The experiment shows that magnetic induction takes place along certain lines. **The directions in the field of a magnet along which magnetic induction takes place are called the lines of magnetic induction, or lines of magnetic force.** They are commonly spoken of simply as **" lines of force."**

Since each piece of iron takes its particular direction on account of the action of the two poles of the magnet upon it, the direction of the curve of the filings at any point represents the direction of the resultant of the forces at that point.

The lines of force can represent not only by their position the direction of the magnetic force, but also by their number its intensity. Just as we speak of measuring the intensity of the illumination of a surface by the number of imaginary rays of light falling upon it, so we speak of estimating the strength of a part of a magnetic field in terms of the number of imaginary lines of magnetic force present in it. But it should be carefully borne in mind that, like the rays of light, the lines of force have no real existence. The actual forces do not act along a set number of lines, but pervade the whole magnetic field.

Experiment 8.

Repeat Experiment 7, placing a suspended magnetic needle in different positions over the card on which the iron filings are placed.

1. **How** does the magnetic needle in its different positions set itself with regard to the direction of the lines of force? Explain.

19

Experiment 9.

Magnetize a needle and suspend it, not by a stirrup, but by a silk fibre tied around it in such a position that it will rest horizontally. Now bring the needle within the field of a bar magnet, placing it at various points around, above, and below the magnet. Remembering that **the magnetic needle always tends to set itself parallel with the lines of force of the magnet,** note the direction of the lines of force at the different points at which the needle is placed (Fig. 196).

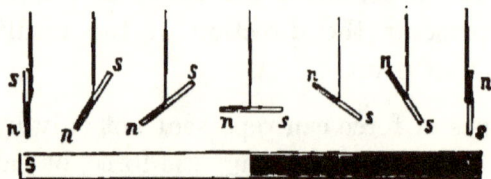

FIG. 196.

1. Do the lines of force all lie within one plane?

2. What is the direction of the lines of force in the **axial line** of the magnet?

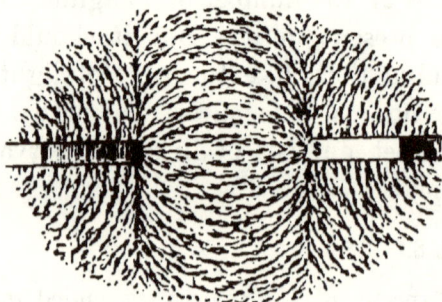

FIG. 107.

17. Superposition of Magnetic Fields.

Experiment 10.

Bring (1) a N-seeking pole of one magnet near a S-seeking pole of another (Fig. 197), (2) a N-seeking pole of one magnet near the N-seeking pole of another (Fig. 198). Place a card over the ends of the magnets in each case, sprinkle iron filings on it, and observe the curves formed when the card is gently tapped.

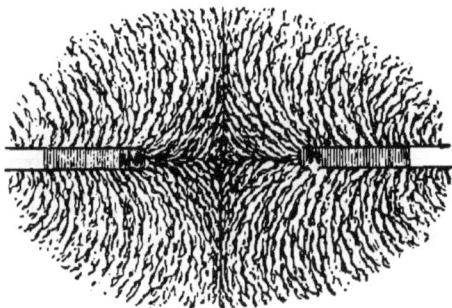

Fig. 198.

Experiment 11.

Repeat Experiment 10, placing the card over (1) a horse-shoe magnet, (2) two bar magnets placed parallel to each other with like poles adjacent and separated from each other about 2 cm., (3) two bar magnets placed parallel to each other and with unlike poles adjacent, (4) a horse-shoe magnet with its keeper on, (5) a bar magnet with a short bar of soft iron near one of its poles, (6) a soft iron ring with one pole of a powerful magnet near it.

1. Make drawings similar to Figs. 197 and 198, indicating the directions of the lines of force in the fields in each case.

2. What is the path of the greater number of the lines of force in (4), (5) and (6), through the air surrounding the poles near the soft iron or through the soft iron itself?

18. Permeability.

Experiment 12.

Interpose between a magnet and an iron tack, (1) a sheet of cardboard, (2) a pane of glass, (3) a thin wooden board, (4) a thick iron plate.

1. Does the magnet attract the tack through each of these bodies?

The magnetic forces act across all bodies which are not themselves magnetic.

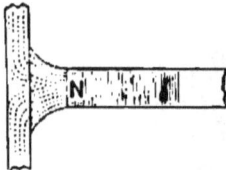

Fig. 190.

The lines of force, instead of passing through a sheet of a magnetic substance into the air on the other side, pass laterally along it (Fig. 199), if it is sufficiently thick, because its **permeability** is many times that of the air.

QUESTIONS.

1. Two similar rods of very soft iron have each of them a long thread fastened to one end, by which they hang vertically side by side. On bringing near the iron rods, from below, one pole of a strong bar magnet, the rods separate from each other. Explain.

2. A bar magnet is held vertically, and two equal straight pieces of soft iron wire hang downward from its lower end. The lower end of each of these wires can by itself hold up a small scrap of iron ; but if the lower ends of both wires touch the same scrap of iron at the same time, they do not hold it up. What is the reason of this?

3. A pole of a magnet is brought within an inch of one side of a sphere of very hard steel, suspended from a string. It manifestly attracts the steel, but is not quite able to draw it into contact. A sphere of iron of the same weight is now substituted for the sphere of steel, and the magnet is found able to draw this new sphere quite up against itself. Explain this difference of action.

4. You have two similar rods, one of steel and the other of soft iron ; you have also a bar magnet and some small iron nails. Describe some experiments which would enable you to distinguish the steel rod from the iron one.

5. If a compass-needle is deflected when a steel bar is brought near it, how can you find out whether the deflection is due to magnetism already possessed by the bar, or to the bar becoming magnetized by the compass-needle at the time of the experiment?

6. A bar magnet is laid upon a table, and a soft iron bar of about the same length as the magnet is hung horizontally just above it by a flexible string. What will be the effect on the soft iron bar if a second bar magnet is laid on the table and brought near the first, at right angles to it, and with its N-seeking pole pointing to the middle of the first magnet? Give a sketch explaining the action.

7. Three precisely similar magnets are placed vertically with their lower ends on a horizontal table. Iron filings are scattered over a plate of glass which rests on their upper ends, two of which are north poles and the third a south pole. Give a diagram showing the forms of the lines of force mapped out by the filings.

8. Why is less force required to pull a small iron rod away from the poles of a powerful horse-shoe magnet than would be required to pull a thick bar of iron away from the poles of the same magnet?

9. A horse-shoe magnet is placed near a compass-needle so as to pull the needle a little way round. On laying a piece of soft iron across the poles of the horse-shoe magnet, the compass-needle moves back toward its natural position. Explain this.

10. A piece of soft iron, placed in contact with both poles of a horse-shoe magnet at the same time, is held on with more than twice the force with which it would be held if it were in contact with only one pole of the same magnet. Why is this?

11. A magnet is placed near a compass-needle so as to pull the needle a little way round. If a thick sheet of soft iron is put between the magnet and the needle, what happens? Why?

12. You have three equal bar magnets without keepers. How would you arrange them so that, when not in use, they might retain their magnetism? Give a sketch.

13. A compass-needle is suspended inside a hollow ball of iron, and an outside magnet will not affect it. Explain.

14. A compass-needle and a straight strip of soft iron of the same length as the compass-needle are fastened together so as to be in contact with each other at both ends. Will the force which tends to make the combination point north and south be the same as that which would act on the compass-needle alone? Give reasons for your answer.

III.—The Earth's Magnetism.

19. The Earth a Magnet.

We have observed in our various experiments that a magnetic needle suspended freely turns always in a definite direction when not influenced by any magnetic substance near it. We can account for this only on the *theory that the earth is a magnet, and that the needle always tends to set itself in the direction of the lines of force of the earth's field which pass through the point at which it is situated.*

20. Declination of the Magnetic Needle.

The magnetic poles of the earth are not coincident with the geographical poles, and hence a suspended magnetic needle does not point exactly north and south. **The angle between the true north and south line and the direction of a compass needle is called the declination of the compass** (Fig. 200). The angle varies in different places, and in the same place is subject to diurnal, annual and other periodic, as well as irregular, variations.

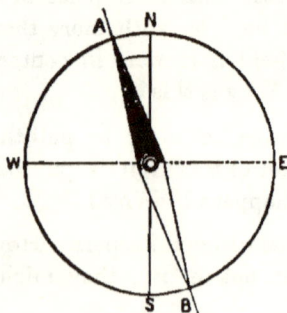

Fig. 200.

21. Inclination, or Dip, of the Magnetic Needle.

Experiment 1.

Take a piece of a knitting-needle and suspend it by tying a silk fibre around it at such a point that it will rest in a horizontal position. Now magnetize it by the method of double touch, and without changing the position of the fibre on the needle, again suspend it.

1. Does the needle now remain in a horizontal position? If not, which pole is turned downward?

The angle between the horizontal and the magnetic axis of a needle which is freely suspended about its centre of gravity is called the inclination, or dip, of the needle.

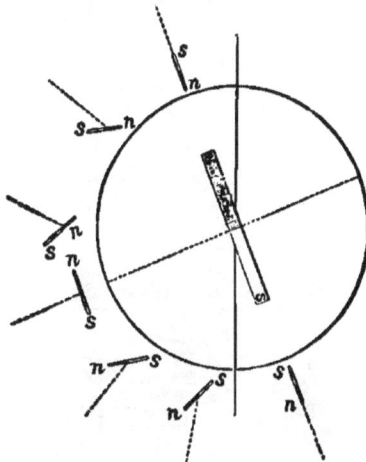

Fig. 201.

The dip of the needle is due to the fact that the lines of force in the earth's field are not horizontal at this latitude.

The lines of force of the earth's field which pass outside the earth run somewhat in the same manner as the lines of force from a bar magnet placed in the centre of a globe (Fig. 201). "N" here represents the pole which is at the north of the earth, and "n" the N-seeking pole of a magnetic needle.

1. Is there any line on the earth's surface where the lines of force of the earth's field are approximately horizontal? If so, where?

2. What change will take place in the dip of the needle as it is carried (1) toward the north pole, (2) toward the equator, (3) from the equator toward the south pole?

3. What is the position of the needle at each of the magnetic poles?

To determine the inclination, a **dipping-needle** is used. Its construction is shown in Fig. 202.

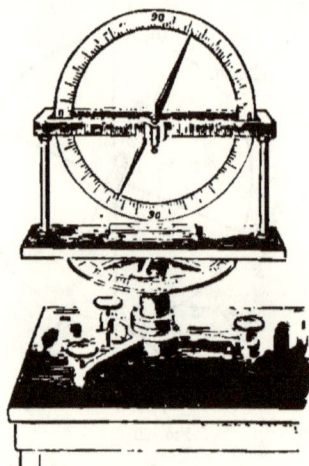

Fig. 202.

The needle is placed in the magnetic meridian, levelled, and the angle of inclination read from the graduated circle placed around it.

22. The Action of the Earth on the Needle Directive, not Attractive.

Experiment 2.

Magnetize a needle or piece of watch spring, and float it on a cork on water.

1. In what position does it set itself?

2. Why does the magnetic attraction not cause the needle and cork to move in a northerly direction?

23. Mariner's Compass.

The mariner's compass consists of a permanent magnet on which is mounted a graduated card, the magnet being so poised on a pivot that the card and magnet remain always horizontal.

FIG. 203.

Fig. 203 shows a common form of the instrument.

QUESTIONS.

1. Given a magnet and the means of suspending it. How will you determine (1) the magnetic meridian, (2) in which direction **North** lies? It is assumed that you do not know which end of your magnet is a N-seeking and which a S-seeking pole.

2. Where on the earth's surface does the N-seeking pole of a magnetic needle point in a generally southerly direction?

3. A tall iron mast is situated a little in front of the compass in a wooden ship. Explain the nature of the compass error when the ship is sailing in an easterly direction (1) in the northern, (2) in the southern hemisphere.

4. If a compass were carried round the equator, would it point in the same direction at all places? If not, state, as nearly as you can, what changes would be observed in its behaviour during the journey.

5. The N-seeking poles of two equal and equally magnetized magnets are attached to the ends of a light bar of wood, so that the magnets are parallel to each other and at right angles to the bar with the S-seeking poles upon opposite sides of it. If the whole is suspended by a thread so that the bar and the magnets lie in a horizontal plane, what position will the bar take up with respect to the magnetic meridian? Give reasons for your answer.

6. Describe the behaviour of a magnetic needle when a bar magnet, with its axis vertical, is moved up and down in its axial line, anywhere in the neighbourhood of the needle.

7. What is meant by saying that the magnetic dip at London is 67° 30'? State in general terms at what places on the earth's surface the magnetic dip is least.

8. The beam of a balance is made of soft iron. When it is placed at right angles to the magnetic meridian two equal weights placed in the opposite pans just balance. Will the weights still appear to be equal when the balance is turned so that the beam swings in the magnetic meridian? Give reasons for your answer.

9. If you were required to make a model to illustrate the magnetic properties of the earth by putting a bar magnet inside a ball of clay, show by a sketch how you would place the magnet, and explain how the magnetic properties of the model would answer to those of the earth.

10. A rod of iron, AB, held in a vertical position with the end B downward, is smartly tapped with a mallet. When turned into a horizontal position and brought near to a compass-needle, the end B repels the north pole of the needle at a distance of four inches, but attracts it when the distance is reduced to one inch. Explain this.

11. A large soft iron rod lies on a table in the magnetic meridian, and a dipping-needle is placed at some distance and at about the same level, (1) due south, (2) due north of it. How will the magnitude of the angle of dip be affected in each case? (Neglect any inductive action between the needle and the bar.)

12. Two bars of soft iron are so placed to the east and west of the N-seeking pole of a compass-needle that the needle still points north and south. If the iron to the east of the needle is replaced by a bar of hard steel of exactly the same size and shape as itself, will the direction in which the magnet points be altered? If so, in what direction will it move, and why?

13. Two equal bars of steel, after having been equally magnetized, are kept for some years in a vertical position, one (a) with its south-seeking pole upward, the other (b) with its north-seeking pole upward. The bars are so far apart that they do not act on each other; which of the two bars would you expect to find had kept its magnetism best, and why?

14. A bar of very soft iron is set vertically. How will its upper and its lower ends respectively affect a compass-needle? Would the result be the same at all points on the earth's surface as at this latitude? If not, state generally how it would differ at different places.

15. Two bars of very soft iron are placed vertically, one east and the other west of the centre of a compass-needle, the lower end of the rod on the east and the upper end of the rod on the west being level with the compass. Describe and explain the effects on the compass.

16. A dipping-needle can oscillate in the magnetic meridian. A long bar of soft iron, held horizontally in a north and south direction, is brought near to it from the south. How is the inclination of the needle to the horizon affected as the distance between it and the bar is gradually diminished?

17. Suppose a magnetic needle to be carried in a circle of a few miles radius round the geographical north pole of the earth. How will the magnetic declination change during one complete circuit? How will it change if the needle is similarly carried round the magnetic pole?

CHAPTER XXV.

THE ELECTRIC CURRENT.

I.—Potential.

1. An Electric Current.

Experiment 1.

Take a strip of zinc about 10 cm. long and 3 cm. wide and connect it with a strip of copper the same size by means of a wire* about 50 cm. or more in length. Fill a tumbler about two-thirds full of water acidulated with about one-twelfth the quantity of sulphuric acid. Place the zinc and copper strips in the acidulated water, not allowing them to touch, and stretch the wire connecting them in a north and south direction (1) over, (2) under a compass-needle (Fig. 204).

Fig. 204.

1. What change takes place in the direction of the needle when the wire is placed (1) over, (2) under the needle?

2. Does this change take place when the wire is above or below the needle, and the strips are removed from the liquid?

---

* Copper magnet wire No. 20 will be found most convenient for making ordinary connections.

The wire evidently possesses new properties when the strips at its terminals are placed in the dilute acid. This result is said to be due to **electricity**.

Just as bodies possess a **thermal condition** called **temperature**, so they are regarded as possessing an **electrical condition**, which is called **potential**; and just as by the transference or the transformation of energy one body may be made to have a higher temperature than another, in consequence of which heat will pass from the one to the other when the two are connected by a thermal conductor, just so it is regarded as possible to cause one body to differ from another in electrical condition, in consequence of which an **"electric current"** will pass from one to the other, when the bodies are connected by an electrical conductor.

The **new properties of the wire** are said to be due to a **current of electricity, which passes through the wire, because a difference in the electrical condition, or potential, between zinc and copper is maintained when they are placed in the dilute acid.**

2. **Potential Defined.**

Potential may be provisionally defined, in general terms, as **that relatively electric condition of a conductor which determines the direction of the transfer of electricity.**

3. **Potential, Temperature, and Level.**

The current of electricity is said to pass from a point of high potential to a point of low potential, as heat passes from a point of high temperature to one of low temperature, or as a liquid at a high level flows to a lower

level; but the analogy between electricity and heat, or
between electricity and a liquid must not be pushed too
far. It certainly is not molecular motion, and hence is
not transferred by an electrical conductor in the same
way that heat is conducted; nor is it matter, as matter ·
is usually defined, because it has no mass that can be
measured.

We are really ignorant of its nature. The theory,
probably, which has most to recommend it for acceptance
is that electricity and ether are identical. If they are
not identical, there is certainly an intimate connection
between them.

4. Positively and Negatively Electrified Bodies.

For the purposes of comparison the earth is taken as a
standard of potential, as the sea is taken as a standard
for the comparison of levels. The earth's surface is
regarded as zero potential, and a body of higher potential
is spoken of as **positively electrified**, and a body of
lower potential as **negatively electrified**. The terms
positive and **negative** are also applied in a general way
to any two related points in a conductor, the **positive**
being **that from which**, and the **negative** that **to which,
the current is flowing**.

5. Different Methods of Causing Bodies to Assume Different Potentials.

Difference in potential between bodies may be brought
about in many ways. A vulcanite rod or a stick of
sealing-wax after being rubbed with flannel differs in
potential from the flannel. The terminals of a Holtz
machine are made to differ in potential when the glass

disc is revolved, and if the difference is sufficiently great a spark passes between them. Under certain conditions of the atmosphere one cloud differs from another, or from bodies on the earth, in potential, and as a result the two clouds, or the cloud and some body on the earth, may be connected by a flash of lightning.

These and similar methods of causing bodies to assume different potentials give rise to an interesting variety of phenomena; but we are concerned only with those methods which tend to keep the ends of a conducting wire at a constant difference of potential, and thus to produce a continuous electric current.

II.—The Voltaic Cell.

6. Simple Voltaic Cell.

Experiment 1, page 300, illustrates one common method of producing an electric current. We shall now repeat the experiment, examining more carefully the **physical and chemical conditions accompanying the generation of the current.**

Experiment 1.

Place the strip of zinc used in Experiment 1, page 300, in the dilute sulphuric acid, and observe its surface for a few minutes.

1. What change takes place in the appearance of the surface of the zinc in the acid ?

2. What is the gas given off ? (See High School Chemistry, page 25.)

Place the copper strip also in the dilute acid and hold it parallel with the zinc and near it, but not touching it (Fig. 205).

1. Does the presence of the copper in any way affect the phenomena observed before ?

2. Does any change take place in the appearance of the surface of the copper ?

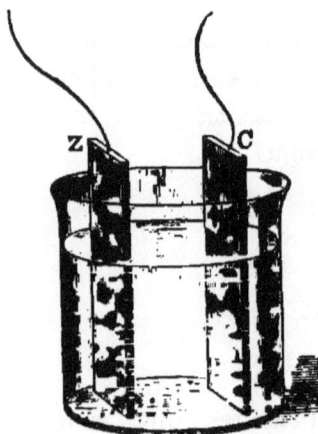
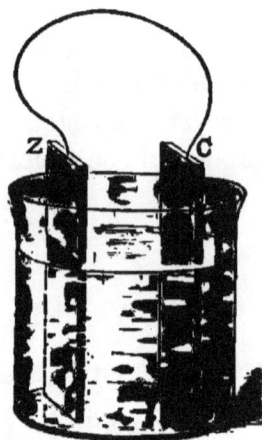

Fig. 205.   Fig. 206.

Connect the upper ends of the strips by a copper wire, or touch them together (Fig. 206).

1. What is now observed at the surface of the copper ?

2. Is there any change at the surface of the zinc ?

3. Is the zinc being acted upon chemically ? To answer this question, weigh the strip, place it in the acid, connect it with the copper, let it stand for a few minutes and weigh it again.

This experiment shows that the production of the electric current in the wire connecting the plates is accompanied by **chemical action.**

The zinc displaces the hydrogen of the acid, forming zinc sulphate, which dissolves in the water; and the liberated hydrogen appears at the surface of the copper plate.

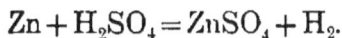

$$Zn + H_2SO_4 = ZnSO_4 + H_2.$$

The arrangement described above is one form of a simple voltaic cell, or element.

The essential parts of any voltaic cell are two different conducting plates immersed in a conducting liquid which acts chemically upon one of them, or if upon both, upon them with unequal power.

In all forms of cells in practical use the plate upon which the liquid acts most powerfully is zinc.

8. Direction of the Current Given by a Voltaic Cell.

The current is regarded as flowing from the surface of the zinc through the liquid to the copper plate, and from it through the external conductor back to the zinc (Fig. 207).

The portion of the zinc plate immersed in the dilute acid is called the positive plate, and the portion of the other plate immersed, the negative plate.

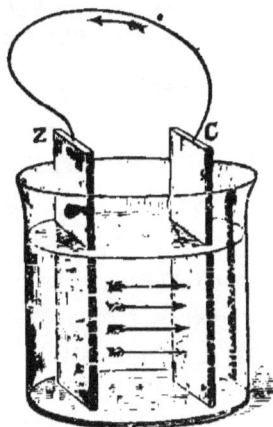

Fig. 207.

The portions of the plates outside of the liquid to which the ends of the external conductor are attached are called the poles of the cell, the external portion of the negative plate being the positive pole, and that of the positive plate the negative pole.

20

9. The Electric Current and the Chemical Action in the Cell.

While the production of an electric current by a voltaic cell is always accompanied by chemical action in the cell, no satisfactory evidence is found to show that the electric current is the result of the chemical action, or that the chemical action is the result of the current. All that can be said is that the one accompanies the other, and that, since the zinc enters into chemical combination, there is a transformation between the energy which produces chemical affinity and the energy of the electric current.

The theoretical reason given for the appearance of the hydrogen on the copper plate will be given when discussing the chemical effects brought about by the current. (See Art. 8, page 335.)

10. To Detect the Presence and the Direction of an Electric Current—The Galvanoscope.

Experiment 2.

Repeat Experiment 1, page 300, holding the connecting wire over the needle in such a way that the current passes (1) from north to south, (2) from south to north.

1. From your observations fill up the proper word, **east** or **west**, in the following rules for determining the direction of a current in a wire:

If a wire is stretched north and south above a magnetic needle and a current is passing from north to south, the N-seeking pole is carried toward the   **; but, if it is passing from south to north, the N-seeking pole is deflected toward the**   **.**

2. Does the same rule apply if the wire is stretched below the magnetic needle? Try.

A feeble current flowing in a single wire either over or under a magnetic needle is unable to deflect the needle; but if the wire is wound into a coil, and the current made to pass several times in the same direction either over or under the needle, it may be deflected. Such an arrangement for detecting the presence of a current is called a **galvanoscope**.

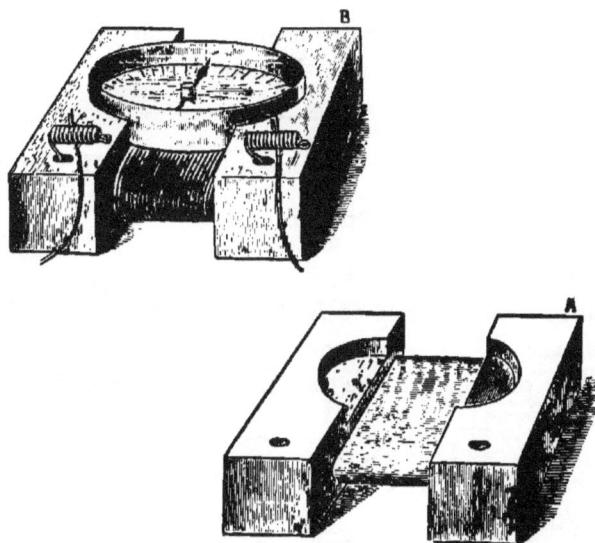

FIG. 208.

11. To Construct a Galvanoscope.

To make a galvanoscope of the form shown in Fig. 208, procure a small compass (the same will answer for the experiments in magnetism), and cut from a piece of board, about three-quarters of an inch thick, a spool of the form A with a recess for holding the compass. The size of the spool will of course depend on the size of the compass used. Fill the spool up to the compass by winding evenly around it cotton or silk-covered magnet wire. Wire of any number between 24 and 30 will answer. Connect the ends of the wire with binding posts. If these are not available, a good substitute may be provided by making a close coil of spring wire as shown in

the figure. The coils are fastened to the ends of the wire
and secured to the wood with tacks or screws. Holes about
one-half inch in diameter should be bored in the spool,
as shown, to serve as mercury cups for making rapid connections. Each of these is connected with the corresponding
binding post by an iron wire.

When in use the instrument must be so placed that the
coils of wire are parallel with the magnetic meridian.

III.—Resistance, Potential-Difference, and Current.

12. Conductors and Non-Conductors.

Experiment 1.

Arrange a voltaic cell as in Experiment 1, page 300, and
connect the wires with the binding posts, as shown in
Fig. 209, placing in the circuit at A (1) a few feet of

Fig. 200.

fine iron wire, (2) the same length of copper wire of the same
size, (3) a string of the same length and about the same size,
(4) a short wooden rod, (5) a glass rod or tube. Observe the
deflection of the needle of the galvanoscope in each case.

1. In which case is the angle of deflection the greatest? In
which cases is the needle not deflected?

The results observed in the above experiment are explained *on the theory that bodies differ in their power to conduct electricity, or in the resistance which they offer to the flow of the current.* When a body is a good conductor of electricity, it offers less resistance to the current than a poor conductor of equal cross section and length; hence if the same difference in potential between the ends is maintained, a stronger current flows through it, and the needle of the galvanoscope is consequently deflected through a greater angle.

The following table gives a list of some of the more common substances classified according to their conductivities:—

Good Conductors	{ Silver. Copper. Gold. Aluminium. Zinc. Platinum. Iron. Tin. Lead. Mercury. Charcoal. Acidulated water.
Partial Conductors	{ The body. Cotton. Dry wood. Marble. Paper.
Non-Conductors	{ Oil. Porcelain. Wool. Silk. Resin. Guttapercha. Shellac. Ebonite. Paraffin. Glass. Air.

Bodies are usually divided into **good conductors, bad** or **partial conductors,** and **non-conductors;** but the distinctions are but relative. Even the best conductor offers some resistance to an electric current, while a weak current of electricity may be made to pass through any so-called non-conductor if a sufficient difference in potential is produced.

A non-conductor used for preventing a current from flowing from one conductor to another is called an **insulator**.

13. Potential Series.

Experiment 2.

Connect the plates of a zinc-copper voltaic cell with the galvanoscope. Note the direction and the amount of the deflection of the needle.

Replace the copper plate by (1) a platinum one, (2) an iron one, (3) a silver one (a silver coin will answer), (4) a carbon one (a piece of an electric light carbon will answer).

1. Is the direction of the deflection of the needle the same n ich case? If so, what does this indicate?

2. Is the strength of the current the same in each case?

3. If not, how can the difference be accounted for?

To answer the last question, consider the case of the zinc-copper and the zinc-platinum cells.

1. When the plates are the same size and kept the same distance apart, which gives the stronger current?

2. Which circuit do you believe will offer the greater resistance to the current?

3. Can you, therefore, account for the difference in current strength by difference in resistance?

4. If not, on what other theory would it be possible to account for it?

Experiment 3.

Connect the plates of a zinc-iron cell with the galvanoscope. Note the direction of the deflection of the needle. Now replace the zinc plate by a carbon one, and note the direction of the deflection of the needle.

1. Does the current now flow in the same direction as before?

2. If not, how can you account for the difference in direction?

The preceding two experiments show that **the difference in potential between different pairs of plates immersed in dilute sulphuric acid is different.** The potential of zinc is higher than iron, and the current flows from the zinc to the iron through the liquid, and from the iron to the zinc through the external circuit. When the zinc is replaced by carbon, the direction of the current is reversed. Thus showing that while zinc when immersed in dilute sulphuric acid is higher in potential than either iron or carbon, iron is higher in potential than carbon under the same conditions.

It is possible by performing experiments similar to the above to arrange different substances in a series in the order of their potentials when immersed in the same exciting fluid. Such a series is called a **potential, or electromotive series.**

Arrange the following substances in order, so that any two being chosen and connected by a conductor a current will flow from the latter to the former through the conductor when they are partially immersed in dilute sulphuric acid: carbon, copper, iron, lead, platinum, silver, tin, zinc.

14. Current and Potential-Difference.

The experiments also indicate that the **strength of the current** passing through the conductor joining the plates **depends not only upon the resistance in the circuit, but also upon the potential-difference between the plates.**

15. Electromotive-Force.

The term **electromotive-force**, or E.M.F., is applied to that which tends to produce a transfer of electricity. In the case of the battery current, the E.M.F. is the result of the potential-difference between the plates. Just as the difference in level in two tanks connected by a pipe causes a pressure which produces a transfer of water through the pipe, so a difference in potential is regarded as producing electromotive force which causes a transfer of electricity through a conductor joining bodies of different potentials; and just as pressure can be estimated in terms of difference of level, for example when we say that the air pressure equals 30 inches of mercury, so electromotive-force may be measured in terms of potential-difference, because it is proportional to it.

The unit electromotive-force is the volt, which is the E.M.F. of a cell of which the potential-difference between the plates is nearly the same as between zinc and copper immersed in diluted sulphuric acid. A more exact definition of the unit will be given at a later stage.

✕   III.   Local Action and Polarization.

16. Local Action.

Experiment 1.

Obtain, if possible, a piece of chemically pure zinc. Immerse it in dilute sulphuric acid. Also immerse in the acid a piece of ordinary commercial zinc.

What difference is observed in the chemical action between the acid and the two pieces of zinc ?

Experiment 2.

Amalgamate a piece of commercial zinc by first dipping it in dilute sulphuric acid to clean it, and then dropping a few drops of mercury on it, and spreading the mercury over its surface by rubbing with a rag or brush.

Immerse the amalgamated zinc in dilute sulphuric acid.

1. Is there any chemical action between the zinc and the acid?

2. Can the amalgamated zinc be used with copper to form a zinc-copper cell?

To answer this question, connect it and a copper plate with the galvanoscope, and partially immerse the plates in dilute sulphuric acid.

1. Does the galvanoscope indicate a current?

2. Does the hydrogen appear as usual at the copper plate?

3. Does the zinc waste away (1) when not connected with the copper plate, (2) when connected with the copper plate? Find out by weighing.

The fact that the commercial zinc wastes away in the dilute sulphuric acid, while the pure zinc and the amalgamated zinc do not, is explained on the theory *that there is a difference in potential between the zinc and its impurities in consequence of which electric currents are set up between the zinc and the impurities in electrical contact with it.* The zinc then enters into combination with the acid when unconnected with any other plate. Such currents are called **local currents**, and the action is called **local action.**

Since a plate of ordinary zinc wastes away in a cell even when unconnected with any other plate without any useful work being done, a plate of amalgamated zinc is commonly used for this purpose. When the zinc is amalgamated the mercury dissolves the pure zinc on the surface, forming a clean uniform layer of pasty zinc amalgam, and the zinc is acted upon by the acid only when it is connected by a conductor with another plate whose potential is different. As the zinc of the amalgam then combines with the acid, the mercury takes up more of the zinc, and the impurities float out into the fluid. Thus a homogeneous surface remains always exposed to the acid.

Why is chemically pure zinc not used in cells instead of amalgamated plates?

17. Polarization.

Experiment 3.

Connect the plates of a zinc-copper cell with the galvanoscope, allow the cell to stand for a few minutes and observe the changes in (1) the appearance of the surface of the copper, (2) the deflection of the needle of the galvanoscope.

1. What evidence have you that the strength of the current becomes weaker the longer the cell stands?

2. What change in the appearance of the copper plate accompanies the weakening of the current?

3. Is there any connection between the change at the surface of the copper plate and the weakening of the current?

To answer this question, remove the copper plate from the liquid, brush off all bubbles and replace it.

Is the current strength increased?

When, through a deposition of a film of hydrogen on the negative plate of a cell, the current becomes feeble, the cell is said to be **polarized.**

18. How Does the Film of Hydrogen on the Copper Plate Cause the Weakening of the Current?

To partially answer this question perform the following experiment:—

Experiment 4.

Place three plates in dilute sulphuric acid, two copper plates, C_1 and C_2, and a zinc plate, Z (Fig. 210). Connect the

FIG. 210.

two copper plates with the galvanoscope, using the mercury cups so that the connections can be made and unmade rapidly.

Does the galvanoscope indicate a current?

Leaving C_1 connected with the galvanoscope, disconnect C_2 and connect the zinc plate with a mercury cup. Allow the cell to stand for a few minutes until C_1 becomes covered with a film of hydrogen, and the current grows weak. Now disconnect the zinc plate and at once connect the two copper plates with the galvanoscope as at first.

1. Does the galvanoscope now indicate a current?

2. If so, does it flow in the same direction as the current given by Z and C_1?

3. Which is of the higher potential, C_1 or C_2?

4. Does the film of hydrogen on the copper plate, therefore, increase or decrease the difference in potential between the copper and zinc plates?

5. What effect will this change in potential have on the strength of the current?

The adhesion of the hydrogen to the copper plate weakens the current in two ways.

1. **By decreasing the potential-difference between the zinc and the copper plate;** because, as was shown in Exp. 4, the copper plate, when covered with the film of hydrogen, is higher in potential than the clean copper plate.

2. **By increasing the internal resistance of the cell,** because the film of gas is a very bad conductor.

19. Methods of Preventing Polarization.

Voltaic cells differ from one another mainly in the remedies provided to prevent polarization. These are numerous, but may be classified as follows:—

1. **By Mechanical means**, that is, by freeing the bubbles of gas from the plate in some mechanical way. Various methods have been proposed, such as keeping the plates in motion, agitating the fluid, etc.; but the only one which has come into practical use is that adopted in the Smee cell.

2. **By Chemical means**, that is, by the use of some substance in the cell which will combine chemically with the hydrogen while it is in the nascent state, and thus prevent its appearance on the negative plate. This is usually a powerful oxidizing agent. The ones most commonly employed are bichromate of potassium, and nitric acid.

This method of preventing polarization is adopted in the following common cells :

Grenet, Grove's, Bunsen's, and Leclanché's.

3. **By Electro-Chemical means**, that is, by employing such plates and such fluids that, not hydrogen, but the same substance as that of which the negative plate is composed, will be deposited on this plate by the action of the current. No change in potential in the plate will then be produced. The negative plate in cells of this class is usually copper. The more common forms are the Gravity cell and Daniell's cell.

IV.—Common Voltaic Cells.

1. Smee's Cell.

Construction.

Fig. 211 shows the construction of Smee's cell. It consists of a silver plate covered with finely divided

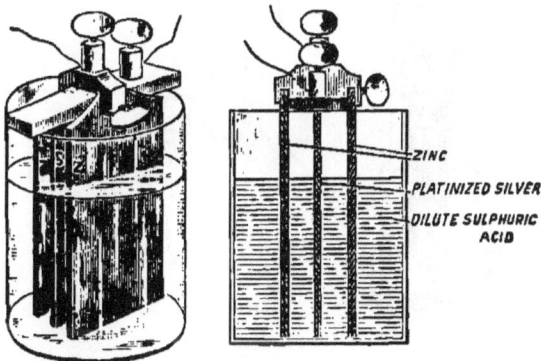

FIG. 211.

platinum, and two connected zinc plates immersed in dilute sulphuric acid.

Chemical Action.

When the plates are connected by a conductor, the zinc displaces the hydrogen of the sulphuric acid, forming zinc sulphate. The liberated hydrogen escapes freely from the numerous sharp points on the platinized silver plate. Polarization is thus prevented for a time.

Current, Etc.

The E.M.F. of the cell is at first nearly one volt, but decreases considerably after a few minutes' use. The current is, therefore, not constant.

2. Grenet or Bichromate Cell.

Construction.

Fig. 212 shows the construction of the Grenet or Bichromate cell. It consists of two connected carbon plates and a zinc plate between them, immersed in a solu-

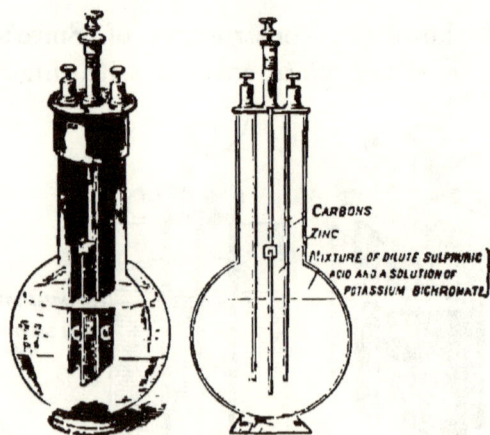

CARBONS
ZINC
MIXTURE OF DILUTE SULPHURIC ACID AND A SOLUTION OF POTASSIUM BICHROMATE

FIG. 212.

tion of potassic bichromate in water mixed with sulphuric acid. The zinc plate is usually attached to a rod, so that it can be raised out of the fluid when not in use.

Chemical Action.

The chemical actions in the cell are somewhat complicated, but the following are the leading ones.

The sulphuric acid acts upon the potassic bichromate,
forming chromic acid. When the circuit is completed,
the zinc displaces the hydrogen of the sulphuric acid,
forming zinc sulphate, and the hydrogen in the nascent
state reduces the chromic acid.   Polarization is thus
prevented.

Current, Etc.

The E.M.F. remains constant at about 2 volts for a
short time, but soon decreases rapidly. The cell is consequently capable of giving a strong current for a few
minutes.

20. To Make a Bichromate Cell.

Fig. 213 shows how a simple form of this cell, with
which the greater number of the effects of the electric
current can be shown, can be
constructed at an expense
of a few cents. Z is a zinc
rod, such as is used in
Leclanché's cell. It may be
obtained from the telephone
company, or from any dealer
in electrical supplies. C, C
are two electric light carbons.
Z and C, C are inserted firmly
into a short bar of wood and
connected by wires as shown

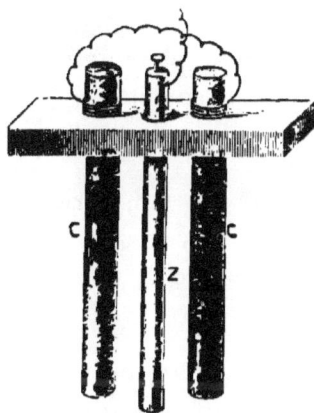

FIG. 213.

in the figure. The fluid may be prepared by dissolving
2 ozs. of potassic bichromate in a pint of water and adding

carefully 2 ozs. by weight of sulphuric acid. The zinc should be kept amalgamated, and the plates should be removed from the fluid when not in use.

3. Grove's Cell.

Construction.

The construction of Grove's cell is shown in Fig. 214. It consists of a zinc plate immersed in dilute sulphuric

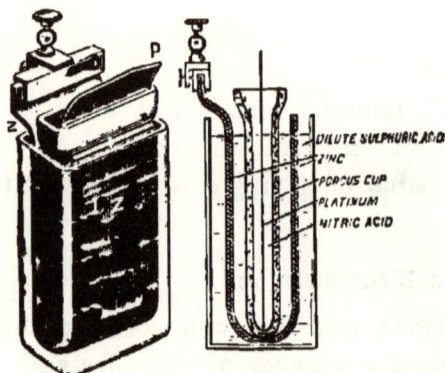

FIG. 214.

acid in an outer vessel, and a platinum plate immersed in nitric acid placed in an inner porous cup.

Chemical Action.

The zinc displaces the hydrogen of the sulphuric acid, forming zinc sulphate, and the nascent hydrogen reduces the nitric acid, thus preventing polarization.

Current, Etc.

The E. M. F. of the cell is about 1.9 volts, and remains nearly constant for some time. One of these cells will furnish an energetic continuous current for three or four hours.

4. Bunsen's Cell.

Bunsen's cell differs from Grove's cell in substituting a carbon plate for a platinum one. Fig. 215 shows a common form of it.

The chemical action is the same as in the Grove cell. The character of the current given by it is also very much the same, the E. M. F. being slightly higher.

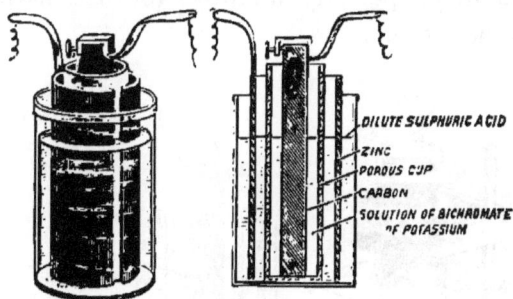

FIG. 215.

5. Leclanche's Cell.

Construction.

The construction of the Leclanché cell is shown in Fig. 216.

It consists of a zinc rod immersed in a solution of ammonic chloride in an outer vessel and a carbon plate surrounded by a mixture of small pieces of carbon and powdered manganese dioxide in an inner porous cup.

Chemical Action.

The solution of ammonic chloride acts upon the zinc, forming a double chloride of zinc and ammonium, and liberating ammonia and hydrogen. The hydrogen is oxidized by the manganese dioxide.

21

Current, Etc.

As the reduction of the manganese dioxide goes on very slowly, the cell soon becomes polarized, but recovers itself when allowed to stand for a few minutes. The E. M. F. is about 1.4 volts at first. As the zinc does not waste away when the circuit is not complete, it does not require renewing for several months, when used intermittently for a minute or two at a time. It is, consequently, specially adapted for use with electric bells, telephones, etc.

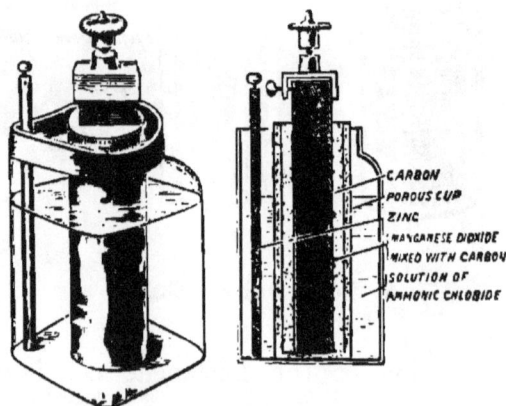

FIG. 216.

6. Gravity Cell.

Construction.

Fig. 217 shows a common form of the cell. A copper plate is placed at the bottom of a vessel, and a zinc plate suspended near the top. Crystals of copper sulphate are placed at the bottom of the vessel around the copper plate, and the vessel is nearly filled with water. The liquid at the lower part of the vessel is, therefore, a saturated solution of copper sulphate.

Chemical Action.

When a little dilute sulphuric acid is added, the zinc displaces the hydrogen of the acid, forming zinc sulphate; the displaced hydrogen in turn displaces the copper of the copper sulphate, forming more sulphuric acid, and copper instead of the hydrogen is deposited on the copper plate. The zinc displaces the hydrogen of the sulphuric acid formed, and so on; hence the zinc wastes away, copper is deposited on the copper plate and zinc sulphate dissolves in the water. The zinc sulphate solution, being much less dense than the copper sulphate solution floats on the top of it, leaving a sharp line of demarkation between the solutions when the cell is undisturbed.

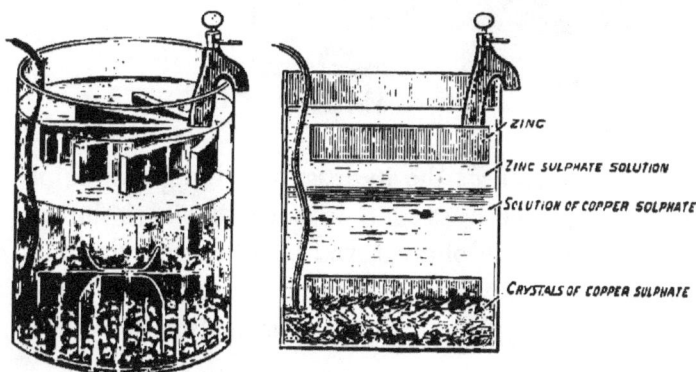

ZINC

ZINC SULPHATE SOLUTION

SOLUTION OF COPPER SOLPHATE

CRYSTALS OF COPPER SULPHATE

FIG. 217.

Current, Etc.

Since nothing but copper is deposited on the copper plate, the cell is never polarized; consequently it is capable of giving a continuous current for an indefinite period, if the materials are renewed at regular intervals. For this reason it is adapted for use with telegraph instruments, or for other closed circuit work.

The E. M. F. is about 1.07 volts.

7. Daniell's Cell.

Fig. 218 shows the construction of the Daniell cell. It consists of a copper plate, which is frequently the outer vessel, immersed in a concentrated solution of copper sulphate, and a zinc plate immersed in dilute sulphuric acid in an inner porous vessel.

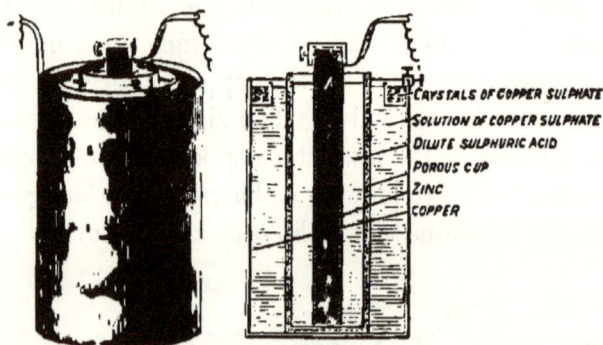

CRYSTALS OF COPPER SULPHATE
SOLUTION OF COPPER SULPHATE
DILUTE SULPHURIC ACID
POROUS CUP
ZINC
COPPER

FIG. 218.

Chemical Action.

The zinc displaces the hydrogen forming zinc sulphate, and the displaced hydrogen in turn displaces the copper from the copper sulphate, forming sulphuric acid; and the displaced copper, instead of the hydrogen, is deposited on the copper plate. Hence polarization is altogether done away with.

Current, Etc.

Since the cell is not subject to polarization, it, like the Gravity cell, may be used for continuous closed circuit work. Its E.M.F. is the same as that of the Gravity cell.

21. Uses of Cells.

Dynamos and storage cells have altogether displaced primary cells whenever a powerful current is required for commercial purposes.

Some form of the Gravity or the Daniell cell is generally employed for working telegraph instruments, but the larger companies are introducing dynamo and storage battery plants for this purpose. The Leclanché cell is used on telephone circuits, and for ringing electric call-bells, gongs, etc.

Bichromate or Bunsen cells are frequently used in the laboratory for illustrating the different effects of the electric current, but storage cells are much more convenient and reliable for this purpose.

While a battery of Bichromate or Bunsen cells may be used as a source of current in all the experiments which follow, we strongly advise the use of a two cell storage battery. It will be found to give less trouble, even where, through the absence of an electric light station, it is found necessary to charge it with Gravity or Daniell cells.

Since storage cells differ widely in efficiency, only those made by reliable companies for actual commercial work should be purchased. Where the battery has to be removed to an electric light station to be charged, the plates should be placed in rubber jars, which should be fitted tightly into a wooden box with convenient handles for moving. The top of the jar should not be sealed. Where the battery can remain stationary in the laboratory, glass jars are the best.

The cells should have a capacity of not less than 50 ampere-hours.

QUESTIONS.

1. You have a Voltaic cell containing a plate of platinum and a plate of zinc immersed in dilute sulphuric acid. What occurs in the liquid when the two metals are united? Does anything occur that you can actually see? If so, what?

2. A steel fork and a steel knife are connected by wires with a galvanoscope. The knife and fork are used to cut a juicy and well-salted beefsteak, what will be the effect upon the galvanoscope? What will be the effect when a silver fork is substituted for the steel one, the steel knife being retained? Why?

3. Place silver on the tongue, and zinc under it, so that the two pieces touch; what occurs? Why?

4. How would the action of a Daniell's cell be modified if the solution of copper sulphate were replaced by dilute sulphuric acid?

5. The platinum plate of a Grove's cell is connected with the copper plate of a Daniell's cell. Would there be a current if the zinc plates were also connected, and if so, in which direction would it flow? What reasons have you for your answer?

CHAPTER XXVI.

THE CHEMICAL EFFECTS OF THE ELECTRIC CURRENT.

I.—Electrolysis.

1. Electrolytes.

Experiment 1.

Connect by means of copper wires, as shown in Fig. 219, (1) one pole of a battery with a strip of platinum, (2) the other pole with one of the binding posts of a galvanoscope, (3) the other binding post of the galvanoscope with another strip of platinum.

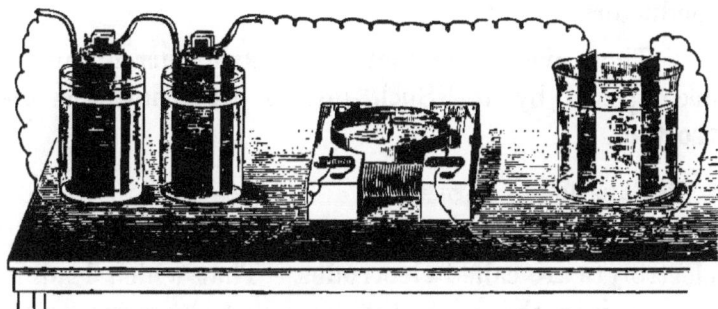

FIG. 219.

Keep the strips from touching each other and dip them into (1) mercury, (2) turpentine, (3) a solution of potassic iodide (KI) to which has been added a little starch paste to serve as a test for iodine (iodine turns starch a deep blue colour).

Observe the deflections of the needle of the galvanoscope, and the appearance of the liquids near the platinum strips.

1. In which cases is the current conducted by the liquids ?

[327]

2. In which does a chemical change take place in the liquid conveying the current? What is the nature of that change?

3. Where are the substances resulting from the chemical changes liberated?

This experiment shows that with regard to their power of conducting an electric current there are three classes of liquids:—

1. Those which, like mercury and molten metals, are conductors, but which are not decomposed by the current.

2. Those, such as turpentine, oils, etc., which are non-conductors.

3. Those which conduct an electric current and are decomposed by it. Such liquids are known as **electrolytes**.

2. Explanation of Terms.

The plates by which the current enters and leaves the electrolyte are called **electrodes**. That which leads the current into the electrolyte is called the **positive electrode**, or **anode**, and that which leads it out the **negative electrode**, or **kathode**.

The process of decomposition by the electric current is called **electrolysis**.

The atoms or groups of atoms which are separated by electrolysis and appear at the electrodes are called **ions**, those appearing at the kathode being called **kations**, and those at the anode **anions**.

The anions are regarded as **electro-negative** and the **kations** as **electro-positive**, because they appear to be attracted to the positive electrode and the negative electrodes respectively.

3. Electrolysis of Water.

Experiment 2.

Arrange apparatus as shown in Fig. 220. The vessel is partially filled with water acidulated with a few drops of sulphuric acid. The test-tubes E and F are also filled with acidulated water and inverted over the platinum strips A and B within the vessel. These strips are then connected by wires C and D with a battery of two or three cells, connected as shown.

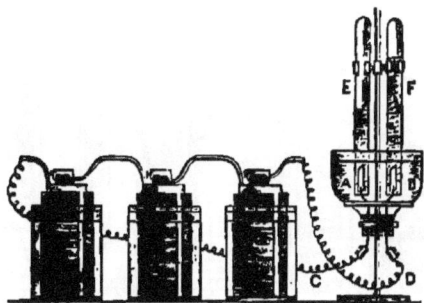

Fig. 220.

1. Describe what takes place.

2. In what proportion by volume are the gases liberated?

3. What are the gases? To answer this question, allow the test-tubes to fill with gas and place a lighted splinter in each.

4. Which is the anion, and which the katione?

4. Electrolysis of Hydrochloric Acid.

Experiment 3.

Arrange apparatus as shown in Fig. 221.

The U-tube is filled nearly full of hydrochloric acid. The wires are passed through the perforated corks, and a small pencil or plate of carbon, to serve as an electrode, is attached to the end of each. The corks are then inserted and the other

ends of the wires connected with the poles of the battery. If
gases are liberated at the electrodes they will pass up through
the glass tubes passing through the corks, and may be collected
in small test-tubes by the displacement of hot water or a saturated solution of common salt.

1. Describe what takes place when the circuit is completed.

Observe the colour of the gases liberated. Test each with a
burning splinter.

2. What is the anion, and what the katione?

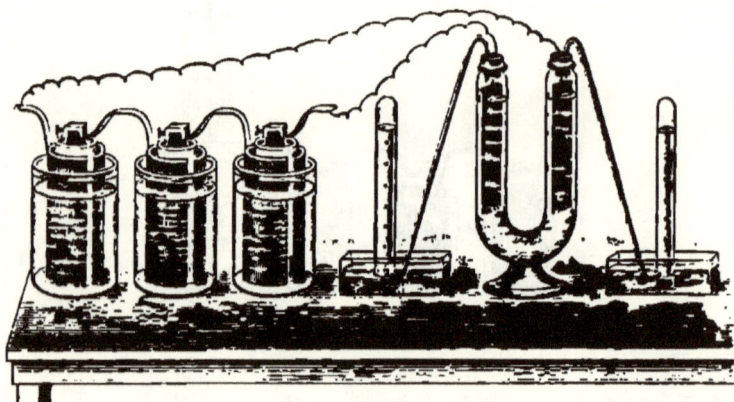

Fig. 221.

5. Electrolysis of Salts.

Experiment 4.

Repeat Experiment 3, attaching platinum strips, instead of
carbon pencils, to the ends of the wires, and filling the U-tube
with a solution of copper sulphate. Collect the gas which is
liberated at one of the electrodes by the displacement of water.

1. What is deposited on the kathode?

2. What gas is liberated at the anode? To answer this question
insert a splinter, at the end of which is a glowing ember, into the
test-tube when filled with the gas.

Experiment 5.

Repeat Experiment 4, allow the current to pass until a deposit is formed on the kathode and then reverse the direction of the current by changing the wires at the poles of the cell.

1. Upon which pole is the deposit now formed?

2. Does the deposit remain on the plate which was at first the kathode?

3. Is gas liberated at either electrode while the change in the deposit is taking place? If not, explain.

Experiment 6.

Weigh two strips of copper, attach each to the pole of a voltaic cell, and, without allowing them to touch, dip them into a solution of copper sulphate.

Is any change observed to take place at either plate? If so, describe it.

When the strips have remained a few minutes in the solution, remove them, and, remembering which was the anode and which the kathode, weigh them again.

1. What change has taken place in the weight of (1) the anode, (2) the kathode?

2. How do you account for these changes?

Experiment 7.

Set up apparatus as for the electrolysis of water (Fig. 220). Fill the vessel and test-tubes with a strong solution of common salt (NaCl), to which has been added sufficient red litmus solution to colour it distinctly.

1. What gases fill the tubes?

2. Account for the change in colour around one of the electrodes.

Experiment 8.

Repeat the last experiment, using a solution of sodium sulphate instead of sodium chloride, and making the litmus purple in colour by exact neutralization.

1. What gases are now liberated ?
2. Account for the change in colour around each electrode.

6. Summary.

The preceding experiments show :

1. That electrolytes are—

   (a) Dilute acids.

   (b) Solutions of metallic salts. Certain fused salts are also capable of electrical decomposition.

2. That when electrolysis takes place the substances resulting from the decomposition of the electrolyte are found at the electrodes. **Hydrogen and the metals are kations, while oxygen, chlorine, iodine,** etc., and **electro-negative radicals are anions.**

3. That in most cases of electrical decomposition, secondary actions, depending on the chemical affinities of the elements involved, take place. For example :—

   (a) In Experiment 4, the radical sulphion (SO_4) which is disengaged from the Cu at the anode, combines with the hydrogen of the water, forming sulphuric acid (H_2SO_4), and liberating oxygen.

   (b) In Experiment 6, the radical sulphion (SO_4) combines with the copper of the anode, forming more of the copper sulphate ($CuSO_4$).

(c) In Experiment 7, the sodium disengaged at the kathode reacts upon the water forming sodium hydroxide and liberating hydrogen.

(d) Even in the case of the electrolysis of water, it is probable that it is the acid which is first decomposed, and that the radical sulphion (SO_4) at the anode combines with the hydrogen of the water, thus liberating the oxygen and forming more of the acid. The quantity of the acid, therefore, remains constant, and the water only is decomposed.

1. What are the chemical changes which take place in Experiment 8?

2. Write equations representing the chemical actions which take place in Experiments 2–7.

7. Theory of Electrolysis—Hypotheses of Grotthuss and Clausius.

The reason for the appearance of the ions at the electrodes and not in the intervening liquid is generally explained by the hypothesis of Grotthuss as modified by Clausius. The theory may be thus stated:

The atoms which form the molecules of the electrolyte are being constantly separated and recombined in the same way, an atom of one molecule displacing a similar one of the next, and it displacing another, and so on. When there is no difference in potential between the electrodes, these atomic changes take place in all directions; but when the plates are of different potentials, the molecules so arrange themselves that the atoms of electro-positive substances are turned toward the kathode and those of electro-negative ones toward the anode, and the exchange of atoms between the molecules takes place in direct lines between the electrodes, the atoms of electro-positive substances moving toward the kathode, and those of electro-negative ones toward the anode. If the electromotive force is sufficiently great, the atoms which thus reach the electrodes remain permanently separated from the electrolyte, and they either combine with one another, or with atoms of other elements present, to form molecules of a kind different from those of the electrolyte.

Figs. 222 and 223 illustrate graphically the above theory.

Fig. 222.

Fig. 222 shows a row of molecules of HCl arranged with the atoms pointing at random in different directions before there is any difference in potential between the plates A and B.

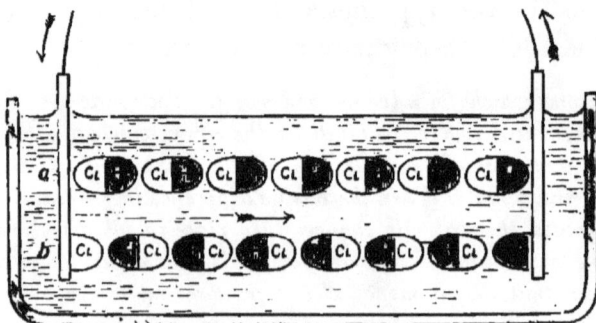

Fig. 223

Fig. 223a shows a row of "polarized molecules' with the atoms of the electro-negative chlorine pointing toward the anode, and those of the electro-positive hydrogen pointing toward the kathode.

Fig. 223*b* shows a row of polarized molecules after an exchange of atoms has taken place, hydrogen being freed at the kathode and chlorine at the anode.

8. Electrolysis in the Voltaic Cell.

According to this theory a similar series of dissociations and recombinations take place in the fluids of the voltaic cell itself. In the simple cell, used in Experiment 1, page 300, the zinc is the anode and the copper the kathode. The electro-negative sulphion group (SO_4) of the H_2SO_4, acts as the chlorine. Zn combines with the SO_4, and the displaced H_2 combines with the SO_4 of the next H_2SO_4 in a direct line toward the copper plate, and the H_2 thus displaced combines with the SO_4 of the next H_2SO_4, and so on. Thus hydrogen appears at the copper plate only, and zinc sulphate is formed at the zinc plate.

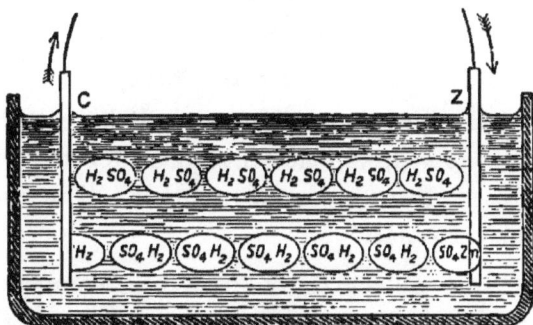

FIG. 224

These changes are shown graphically in Fig. 224.

When a depolarizing fluid, for example copper sulphate, is used, the displaced H_2 combines the SO_4 of the

CuSO$_4$ next it, and the displaced Cu combines with the SO$_4$ of the next CuSO$_4$, and so on. Thus finally copper is deposited on the copper plate.

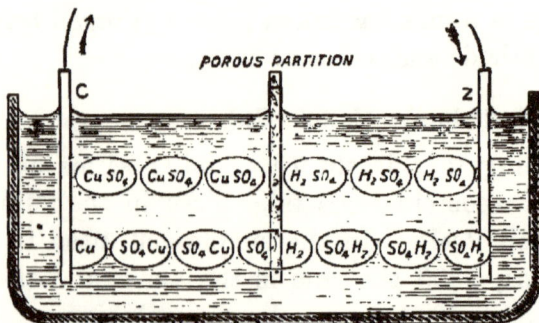

POROUS PARTITION

FIG. 225.

Fig. 225 represents graphically this action.

Trace the intermolecular atomic changes which are supposed to take place in a Bunsen cell when nitric acid is the depolarizing fluid.

II.—Practical Applications of the Chemical Effect of the Current.

1. Electroplating.

The deposition of a metal from a salt by means of an electric current is taken advantage of for covering one metal with a thin layer of another. The process is known as electroplating.

The metallic object to be plated is connected by a conductor with the negative pole of a battery or dynamo, and immersed in a bath containing a solution of a salt of the metal with which it is to be plated. A plate of this metal is also immersed in the bath and is connected by a conductor with the positive pole of the battery or dynamo; that is, the object to be plated is made the kathode, the metal with which it is to be plated is made the anode, and the electrolyte is a salt of this metal. When the current passes through the solution from the plate to the object, the salt is decomposed and the metal is deposited on the object; but as the radical of the salt combines with the metal forming the anode, the strength of the solution remains constant. The metal is thus transferred from the plate to the object.

For copper plating, the bath is usually a solution of copper sulphate; for gold and silver plating, a solution of cyanides of these metals are commonly used.

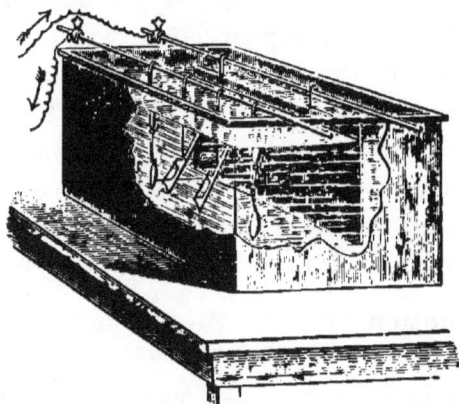

Fig. 226.

Fig. 226 shows a bath and the connections for silver plating.

2. Electrotyping.

Books are now usually printed from electrotype plates instead of from type. These are made as follows:—

An impression of the type is made in a wax mould. This is covered with powdered plumbago to provide a

22

conducting surface upon which the metal can be deposited. The mould is flowed with a solution of copper
sulphate, and iron filings are sprinkled over it. The
iron displaces copper from the sulphate, and the
plumbago surface is thus covered with a thin film of
copper. The iron filings are washed off, and the mould
immersed in a bath of nearly concentrated copper sulphate solution slightly acidulated with sulphuric acid.
The copper surface is then connected by a conductor with
the negative pole of a battery or dynamo, and a copper
plate which is connected with the positive pole is immersed in the bath.

When the current passes, the copper sulphate is decomposed and a layer of copper is deposited uniformly on
the mould, while the copper anode combines with the
sulphion (SO_4) groups to form more of the copper sulphate. When the layer of copper has become sufficiently
thick it is removed from the bath, backed with melted
type-metal and mounted on a wooden block. The face
is an exact reproduction of the type or engraving.

3. Reduction of Ores—Electricity Applied in Manufactures.

Electrical decomposition is sometimes resorted to for
reducing metals from their ores. A soluble or fusible
salt is formed by the action of chemical reagents, and the
metal is deposited from this by electrolysis.

A current of electricity is now frequently employed
for preparing chemical products for commercial purposes.
For example, white lead (lead carbonate) is now made by
causing a current of electricity to pass between two
lead plates immersed in dilute nitric acid, while a stream

of carbon dioxide is passed through the solution. A current of electricity is also used in the manufacture of potassium chlorate from potassium chloride. Electricity is just beginning to be applied to such purposes, and electrical processes for the production of useful compounds are likely to become much more general.

4. Secondary or Storage Cells.

9. Polarization of Electrodes.

Experiment 1.

Connect by means of wires two platinum strips with the poles of two Bichromate or Bunsen cells, placing a galvanoscope in the circuit. Keep the strips from touching, and immerse them in water acidulated with sulphuric acid. Observe

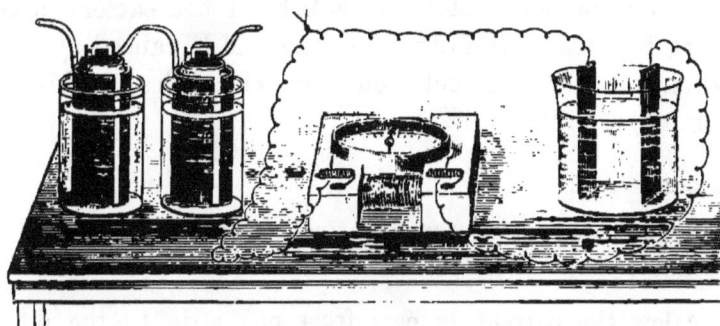

FIG. 227.

the direction of the deflection of the needle of the galvanoscope, and as soon as the gases are given off freely from the strips, disconnect the wires from the poles of the battery and at once join them together, as shown in Fig. 227.

1. What evidence have you that a current of electricity flows through the wires when they are disconnected from the battery and joined ?

2. Does the current flow in the same direction as the battery current, or in a direction opposite to it ?

3. How long does the current continue to flow ?

When a film of hydrogen surrounds one platinum strip in the dilute sulphuric acid and a film of oxygen the other, there is a difference in potential between the strips, which causes a current to flow from one to the other when they are joined by a conductor. The electrodes are then said to be **polarized.**

As the hydrogen is of higher potential than the oxygen, the direction of the current in the conductor joining the electrodes will be opposite to the direction of the current which deposited the oxygen and hydrogen. Hence, in order to overcome this difference in potential and to decompose water, the E.M.F. of the battery used must be greater than the opposite E.M.F. caused by this potential-difference between the electrodes. This is about 1.47 volts.

Experiment 2.

Repeat the last experiment, using two lead strips instead of platinum ones. They should be an inch or more in width.

Allow the current to pass from one strip to the other through the dilute acid for a few minutes. Observe the direction of the deflection of the needle of the galvanoscope and any changes which take place in the appearance of the surface of either strip. Disconnect the wires from the poles of the battery, and join their ends as in the last experiment. Again observe the direction of the deflection of the needle of the galvanoscope, and any changes in the appearance of the surface of either strip of lead.

1. What changes are observed to take place in either lead strip (1) when the battery is in the circuit, (2) when the battery is disconnected and the ends of the wires joined ?

2. What is the cause of the current which flows through the wires when the battery is disconnected and the circuit completed ?

3. How does the direction of the battery current compare with that given by the lead strips immersed in the dilute acid ?

4. Can the latter current be used to ring an electric bell ? Try.

10. Secondary or Storage Cells.

The last experiment illustrates the principle of action of all secondary, or storage, cells.

When the current is passed through the dilute acid from one plate to the other the oxygen freed at the anode unites with the lead, forming an oxide of lead. The composition of the anode is thus made to differ from the kathode, and in consequence there is a difference in potential between them, which causes a current to flow in the opposite direction when the plates are joined by a conductor.

This current will continue to flow until the plates become again alike in composition, and hence in potential.

Instead of using solid lead plates, perforated plates, or "grids," made of lead or some alloy of lead, are frequently employed. The holes in the positive plates are filled with a paste made of red lead (Pb_3O_4) and sulphuric acid, and those in the nega-

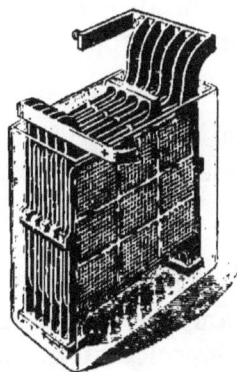

Fig. 228.

tive plate with a paste of litharge (Pb_2O_3) and sulphuric acid. (Fig. 228.)

When the plates are immersed in dilute sulphuric acid, and the current passed from the positive to the negative plates, the positive ones are oxidized to a higher oxide while the negative ones are reduced by the hydrogen liberated. The cell is then said to be charged. During discharge the opposite chemical action takes place.

1. What transformations of energy take place in (1) charging a secondary cell, (2) discharging it

2. Is anything "stored up" in a storage cell? If so, what?

5. Measurement of the Current—Voltameters.

11. Laws of Electrolysis.

Carefully repeated quantitative experiments have verified the following laws of electrolysis.

Law I.—**The amount of an ion liberated at an electrode in a given time is proportional to the strength of the current.**

Law II.—**The weights of the elements separated from an electrolyte by the same electric current are in the proportion of their chemical equivalents.**

These laws furnish a means of comparing the strength of one electric current with that of another, and hence of measuring a current when a unit current is adopted.

12. Unit Current.

The practical unit of current commonly adopted is the **ampere, which may be defined to be a current which deposits silver at the rate of 0.001118 grams per second.** The same current deposits per second 0.000328 grams of copper, and liberates 0.000010386 grams or 0.1168 c.cm of hydrogen.

The weight of an element liberated in one second by a current of one ampere is called the **electro-chemical equivalent** of the element.

An electrolytic cell used for the purpose of comparing the strengths of different currents is called a **voltameter**.

13. Silver Voltameter.

The silver voltameter consists of a light platinum bowl partially filled with a solution of silver nitrate in which is suspended a silver disc. When the voltameter is placed in the circuit, the platinum bowl is made the kathode, the silver disc the anode, and the current to be measured is passed through the silver nitrate solution for a specified time. The silver disc is then removed, the solution of nitrate poured off, and the silver deposited in the bottom of the bowl washed, dried, and weighed. The rate in grams per second, at which it is deposited is then calculated, and this divided by 0.001118 gives the measure of the current in amperes, or

$$C = \frac{W}{t \times .001118}$$

when C is the measure of the current in amperes, and W is the weight of silver deposited by it in t seconds.

The current should not exceed one ampere for each six square inches of the surface of the electrodes.

14. Copper Voltameter.

The copper voltameter consists of two copper electrodes immersed in a solution of copper sulphate. The plate made the kathode is weighed, and the current to be measured is passed through the copper sulphate solution for a specified time. The kathode is then removed, washed, dried, and weighed. If W grams is the increase in weight in the kathode in t seconds,

$$C = \frac{W}{t \times .000328}$$

where C is the measure of the current in amperes.

The surface of each plate immersed in the copper sul-

Fig. 229.

phate solution should be at least two square inches for each ampere of current to be measured. Fig. 229 shows a common form of the instrument.

15. Water Voltameter.

The water, or hydrogen, voltameter consists of the apparatus used for the decomposition of water (Experiment 2, page 329). The vessel is filled with water acidulated with a few drops of sulphuric acid. A graduated tube is also filled with the water, and is inverted over the platinum strip made the kathode.

The current to be measured is passed through the water until the liquid in the tube stands on a level with the liquid in the vessel, and the time during which the current is passing is noted. The temperature of the gas and the barometric pressure are also noted. The volume of the hydrogen liberated is read from the graduated tube, reduced to standard temperature and pressure, and the mass corresponding to this volume calculated.

Then, if the current is passing for t seconds, and W grams is the weight of the hydrogen liberated,

$$C = \frac{W}{t \times .000010386}$$

where C is the measure of the current in amperes.

QUESTIONS.

1. Can a single Gravity cell be used to decompose water ? If not, why ?

2. Give an example of (1) chemical combination, (2) chemical decomposition, brought about by the same voltaic current.

3. When a plate of zinc and a plate of platinum connected by a wire are both dipped into the same vessel of dilute sulphuric acid, an electric current passes through the wire. State and account for the effect of moving one of the plates into a separate vessel of acid.

4. Two copper wires, one connected with one terminal of a voltaic battery and the other connected with the other terminal, dip side by side, but without touching each other, into a solution of sulphate of copper. What happens to the immersed part of each wire ?

5. Plates of copper and platinum are dipped into a solution of copper sulphate, and a current is passed through the cell from the copper to the platinum. Describe the effects produced ; also what happens when the current is reversed.

6. A vertical partition of porous earthenware is fitted into a tumbler, and dilute sulphuric acid is poured into each compartment. Rods of common zinc and copper are placed respectively in the two compartments, and connected by a wire. State what will be observed with regard to the evolution of gas, and how the observed phenomena will be modified when copper sulphate is poured into the compartment containing the copper rod.

7. A piece of zinc and a piece of copper are each carefully weighed ; they are then connected by a copper wire and dipped side by side into dilute sulphuric acid contained in an earthenware jar. After, say half an hour, the pieces of zinc and copper are taken out of the acid, washed and dried, and weighed again. Would the weights be the same as at first ? If not, how, and why, would they differ ?

8. A vessel containing a solution of salt, coloured with a little litmus, or indigo, is divided into two parts by a partition formed by stitching together several layers of blotting paper. The wires coming from the poles of a Grove's battery are dipped into the liquid on opposite sides of this partition. On one side the colour is observed to disappear. Explain its disappearance, and mention the pole of the battery from which the wire that destroys the colour proceeds.

9. The same current is passed through three electrolytic cells, the first containing acidulated water, the second a solution of copper sulphate, and the third a solution of silver nitrate. What weight of hydrogen and oxygen will be liberated in the first cell, and what weight of copper deposited on the kathode of the second cell when 11.18 grams of silver are deposited on the kathode of the third cell?

10. Is there polarization of the electrodes in (1) the water voltameter, (2) the copper voltameter, (3) the silver voltameter ? Give reasons for your answer.

11. Obtain two copper plates, make a copper voltameter, and measure with it the current given by any cell.

CHAPTER XXVII.

THE MAGNETIC EFFECTS OF THE CURRENT.

I.—Electricity and Magnetism.

1. Magnetic Field Due to an Electric Current.

Experiment 1.

Pass a strong current from a battery* through a copper wire, dip the wire into iron filings, and lift it out.

1. What is observed?

Break the circuit.

2. What now takes place? Why?

Experiment 2.

Pass a thick wire vertically through a hole in the centre of a card. Sprinkle iron filings from a muslin bag over the card, Fig. 230. Now connect the ends of the wire with the poles of a battery, and gently tap the card.

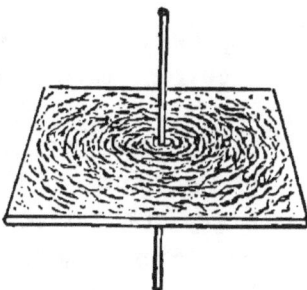

Fig. 230.

1. How do the iron filings arrange themselves around the wire?

2. What does this prove?

---

* If a storage battery is used for experiments of this class, care should be taken to keep from "short-circuiting" it, that is, using it with a resistance so low that the battery discharges at too high a rate. To prevent this, a resistance coil should be permanently attached to one of the poles of the battery. A suitable coil of iron telegraph wire will answer well. The rate of discharge will depend upon the number and the size of the plates in the cell. The rate should be ascertained from the maker.

Experiment 3.

Repeat Experiment 1, page 300.

The above experiments show that **a wire through which an electric current is flowing is surrounded by a magnetic field, the lines of force of which pass in circles around it**; that is, the wire throughout its whole length is surrounded by a "sort of enveloping magnetic whirl." The poles of a magnetic needle placed in this field are apparently urged with equal force in opposite directions around the wire, and it therefore remains at a tangent to it.

Experiment 3 shows that the direction in which each pole of the magnetic needle tends to turn around the wire depends on the direction of the current. **If we imagine a current to flow through a wire from an observer to the face of a clock, the N-seeking pole of a magnetic needle placed in its field tends to turn in the direction of the hands of the clock, while the S-seeking pole is urged in the opposite direction.** If there were but one pole to the magnet it would apparently revolve around the wire continuously.

2. The Magnet and the Solenoid.

Experiment 4.

Make a helix, or coil, of wire 2 or 3 inches long by winding insulated copper wire No. 20 around a lead pencil. Connect the ends of the wire to the poles of a battery, and pass a magnetic needle around the coil.

1. How does the magnetic needle set itself when placed (1) near each end of the spiral, (2) midway between the ends?

2. In what particulars does the helix resemble a bar magnet?

3. What pole of the helix is the observer in front of when the current in the coils facing him is passing in the direction of the hands of a clock?

Experiment 5.

Solder one end of the wire of a helix similar to that used in the last experiment to a small zinc plate, and the other end to a similar copper plate, and pass the plate through a large cork, as shown in Fig. 231. Float the cork in dilute sulphuric acid contained in a large glass basin.

In what direction does the helix set itself when undisturbed?

Present one pole of a bar magnet, first to one end, and then the other, of the helix.

What takes place?

Fig. 231.

Pass an electric current from a battery through another similar helix of wire, and present one end of this, first to one end, and then to the other, of the helix supported by the cork.

1. What now takes place?

2. What properties does a helix of wire through which an electric current is passing possess?

Experiment 6.

Make a helix of insulated wire, No. 16 or 18, about $\frac{3}{4}$ inch in diameter and 3 inches long, and place it in a rectangular

opening made in a sheet of cardboard, so that its axis will be
in the plane of the cardboard (Fig. 232). This can be done
by cutting out the three sides of a rectangle of the proper size,
and then passing the free end of the strip through the centre
of the helix, and replacing the strip in position. Sprinkle
iron filings from a muslin bag on the cardboard around the
helix and within it. Attach the ends of the wire to the poles
of a battery, and gently tap the cardboard.

How do you account for the way in which the iron filings arrange
themselves ?

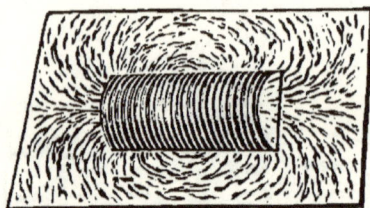

Fig. 232.

The above experiments show that **a helix of wire
through which an electric current is passing acts
exactly like a magnet, having two poles and a neutral
equatorial region.** The field which surrounds it resembles that of a bar magnet.

Such a coil is sometimes called a **solenoid.**

3. Electro-Magnets.

Experiment 7.

Repeat Experiment 4, passing a small soft iron rod through
the helix before the current is passed through the wire.

1. What effect has the introduction of the iron upon the magnetic
power of the helix ?

2. Are the N-seeking and S-seeking poles at the same ends of
the helix as before the insertion of the core ?

Will the end of the rod lift up a small piece of iron, such as a tack, (1) when the current is passing through the wire, (2) when the circuit is not completed ?

A soft iron core surrounded by a helix of insulated wire, through which an electric current can be passed, is called an **electro-magnet.**

Why is an electro-magnet a more powerful magnet than a solenoid ?

To answer this question repeat Experiment 6, placing a soft iron core in a helix of wire.

1. How does the arrangement of the iron filings on the card differ from that observed when the core was not inserted ?

When the helix is used without the core the greater number of the lines of forces pass in circles around the individual turns of wire, and but a few run through the helix from end to end, and back again outside the coil; but when the iron core is inserted the greater number of the lines of force pass in this way, because the permeability of iron is very much greater than that of air, and whenever a coil passes near the core, the lines of force, instead of passing in closed curves around the wire, change their shape and pass from end to end of the core. The effect of the core, therefore, is to increase the number of lines of force which are concentrated at definite poles, and consequently to increase the power of the magnet.

4. Polarity of an Electro-Magnet and Direction of the Current.

Looking at the S-seeking pole of an electro-magnet, the magnetizing current is passing through the coils in the direction of the hands of a clock, and, looking

at the N-seeking pole, the current is circulating in the opposite direction (Fig. 233).

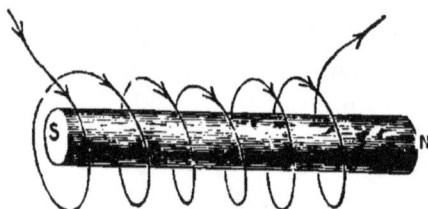

Fig. 233.

5. Use of Solenoid.

Experiment 8.

Make a solenoid about 4 inches long by winding four or five layers of No. 20 insulated wire around a glass or cardboard tube. Connect the ends of the wire with a battery, hold the tube in a vertical position, and take a short soft iron rod which will just slip easily into the bore of the tube, and insert it part way into the tube.

How does the rod tend to set itself within the helix?

A solenoid with a movable iron plunger is frequently used instead of an electro-magnet with a permanent core, when the magnet is required to give a pull through a long range.

6. Laws of Magnets.

Experiment 9.

Take a soft iron rod, $1\frac{1}{2}$ inches in diameter and 2 or 3 inches long, and wind around it one layer of insulated wire, No. 20. Connect the ends of the wire with the poles of a battery, and test the lifting power of the magnet by trying to lift small pieces of iron with .it. Repeat the experiment, winding two, three, four, etc., layers of wire on the rod.

1. What effect has increasing the number of layers of wire upon the power of the magnet? Why?

2. If the same difference in potential is always maintained between the ends of the wire, will the power of the magnet always continue to be affected in the same way by increasing the number of turns of wire? If not, why?

3. If the same current is maintained in the wire, will the power of the magnet always continue to be affected in the same way by increasing the number of turns of wire? Give reasons for your answer.

Experiment 10.

Connect in a circuit with a battery an electro-magnet and a rheostat, or series of resistance coils. Test the lifting power of the magnet. By lessening the number of the coils of the rheostat in the circuit, increase the current. Again test the lifting power of the magnet.

What effect has increasing the current on the lifting power of the magnet?

Repeat the experiment, decreasing the current by increasing the resistance.

1. What change now takes place in the strength of the magnet?

2. What is the relation between the current and the strength of the magnet? Why?

These experiments illustrate the following laws:

Laws of Magnets.

1. **The strength of an electro-magnet is proportional to the strength of the current.**

2. **The strength of an electro-magnet is proportional to the number of turns of wire, if the current is kept constant.**

23

These laws are true only when the iron core is not near the point of being magnetized to saturation.

It should also be observed that when an electro-magnet is used with a battery, or other source of current where the ends of the wire are kept at a constant difference in potential, an increase in the number of turns of the wire may not necessarily add to the strength of the magnet, because the loss in power through loss in current caused by increased resistance may more than counterbalance the gain through the increased number of turns of wire.

In what circuit should a "long coil" electro-magnet (one with a great number of turns of fine wire) be used, one in which the remaining resistance is great or small as compared with the resistance of the magnet?

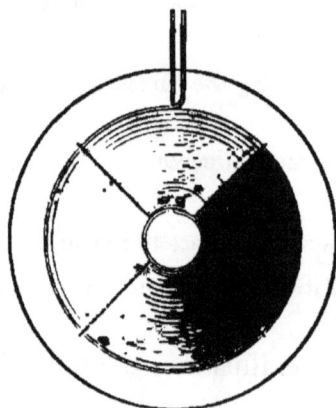

FIG. 234.

7. Laws of Currents.

Law I.

Experiment 11.

Make two coils of wire, of the form shown in Fig. 234, by winding insulated magnet wire No. 20 on a piece of cardboard

and stitching it to the board.   Each coil should be about
3 inches in diameter.

When the coil is finished, allow the free wires to stand out
8 or 10 inches from the coil, bend a hook in the end of each,
and suspend the coils from metallic eyes attached to binding
posts in such a way that they rest facing, and at a distance of

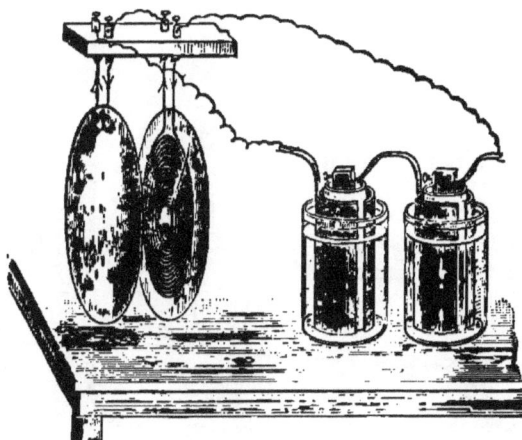

FIG. 235.

not more than a quarter of an inch from each other (Fig. 235).
Connect the binding posts with a battery so that the current
will pass (1) in the same direction through the coils, (2) in
opposite directions.

What happens in each case ?

**Parallel currents in the same directions attract each other;
parallel currents in opposite directions repel each other.**

When the currents flow in the same direction, their
magnetic fields tend to merge, and the stress in the
medium which surrounds the wires tends to draw them
together, but when the currents flow in opposite directions the stresses tend to push the wires further apart.

To show the directions of the lines of force in the fields repeat Experiment 2, page 347, passing two wires through the card and causing the current to pass (1) in the same direction through each wire (Fig. 236), (2) in opposite directions (Fig. 237).

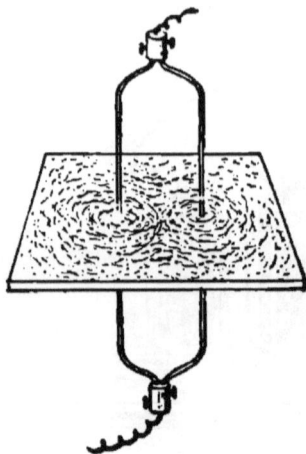

Fig. 236.

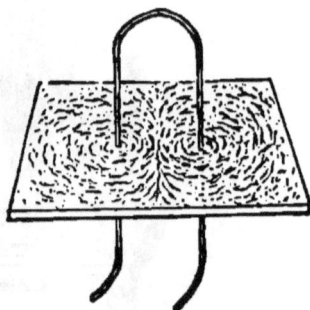

Fig. 237.

Law II.

Experiment 12.

Arrange the floating battery with the solenoid, as described in Experiment 5, page 349. Allow it to stand undisturbed, and stretch a wire through which an electric current is flowing over the solenoid and parallel with it (Fig. 238). Trace the direction in which each current is flowing, and observe the change in the position of the solenoid when the wire is held above it.

What evidence have you that the two currents tend to flow in parallel planes and in the same direction?

Angular currents tend to become parallel and to flow in the same direction.

Show how this law results from the doctrine of stresses in the medium surrounding the wires.

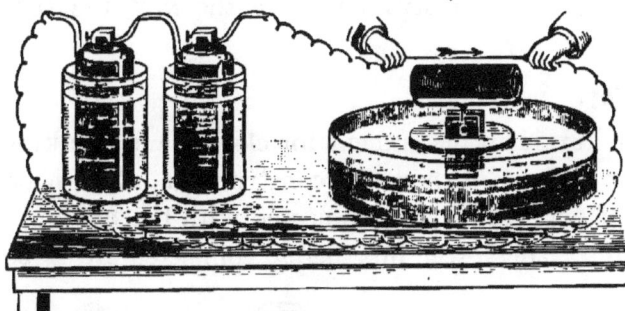

FIG. 238.

8. Ampere's Theory of Magnetism.

Ampère, perceiving that a solenoid acts in every way like a magnet, offered as an explanation of the magnetization of magnetic bodies the theory that *the molecules had closed electric currents circulating around them, thereby causing them to exhibit polarity. During the process of magnetization the currents are caused to become parallel, and the magnetic axes of the molecules are thus made to lie in one direction.*

The currents, like the molecules themselves, can neither be produced nor destroyed, and their permanence can be accounted for only on the theory that they meet with no resistance, because in this way only can their energy be dissipated.

II.—Practical Applications of the Magnetic Effects of the Current.

1. The Electric Telegraph.

The telegraph instruments are the **key**, the **sounder** and the **relay**.

The Key.

The key is an instrument for closing and breaking the circuit.   Fig. 239 shows its construction.   Two platinum contact points P, P, are connected with the binding posts A and B, the lower one being connected by a bolt C

Fig. 239.

insulated from the frame, and the upper being mounted on the lever L which is connected with the binding post B by means of the frame.   The key is placed in the circuit by connecting the ends of the wire to the binding posts.

When the lever is pressed down the platinum points are brought into contact and the circuit is completed. When the lever is not depressed a spring N, keeps the points apart.   A switch S, is used to connect the binding posts, and close the circuit when the instrument is not in use.

The Sounder.

Fig. 240 shows the construction of the sounder. It consists of an electro-magnet, E, above the poles of which is a soft iron armature A mounted on a pivoted beam B, the beam being raised and the armature held by a spring S above the poles of the magnet at a distance regulated by the screws C and D. The ends of the wire of the magnet are connected with the binding posts.

Fig. 240.

The Relay.

The relay is an instrument for closing automatically a local circuit in an office when the current in the main circuit, on account of the great resistance in the line, is too weak to work the sounder. It is a key worked by an electro-magnet instead of by hand. Fig. 241 shows its construction. It consists of a "long coil" electro-magnet R, in front of the poles of which is a pivoted lever L carrying a soft iron armature, which is held a little distance from the poles by a spring S. Platinum contact points P, P, are connected with the lever L and the screw C. The ends of the wire of the electro-magnet are connected with binding posts B, B, and the lever L and screw C are electrically connected with the binding posts B_1, B_1.

Whenever the magnet E is magnetized the armature is drawn toward the poles and the contact points P, P, are brought together and the local circuit completed.

Why should the magnet in this instrument be a "long coil" magnet?

FIG. 241.

Fig. 242 shows a telegraph line passing through three offices, A, B and C, and indicates how the connections are made in each office.

Action.

When the line is not in use the switch on each key is closed, and the current in the main circuit flows from the copper plate connected with the wire at the main battery at A through the wire, across the switches of the keys, and through the electro-magnets of the relays, to the first zinc plate of the main battery at C, and from the zinc plates to the copper plates through the fluid of the battery, and thence through the wire to the ground, which forms the return circuit to the zinc plate of the main battery at A. The magnets R, R, R, are magnetized, the local circuits completed by the relays, and the current from each local battery flows through the

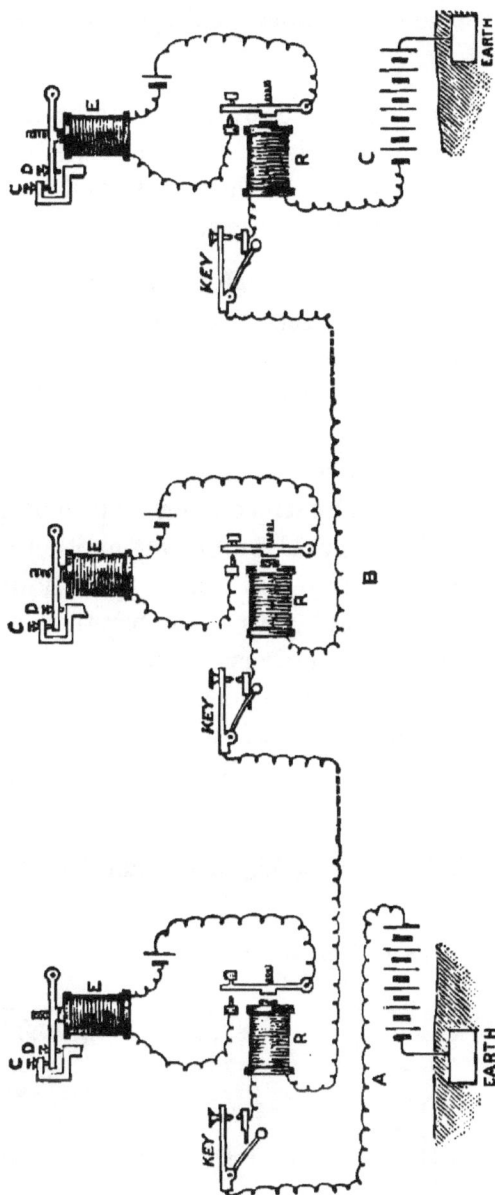

FIG. 242.

magnet E of the sounder, thus holding the screw D against the frame.

When the line is to be used by an operator in any office A, the switch of the key is opened, the circuit broken, and the armature of the relay and of the sounder in each of the offices released.

When he depresses the key and completes the main circuit, the armature of the relay in each office is drawn in, the local circuit is completed, and the screw D of each sounder is drawn down against the frame, producing a click. When he breaks the circuit at the key, the armature of the relay in each office is released, the local circuit is broken, and the beam of each sounder is drawn up by the spring against the screw C, producing another click. When the circuit is completed and broken quickly by the operator, the two clicks are very close together, and a "dot" is formed; but when an interval intervenes between the clicks the effect is called a "dash." Different combinations of "dots" and "dashes" form different letters. The operator is thus able to make himself understood by the listener.

The following is the code of signals generally adopted in America:

MORSE CODE OF SIGNALS.

A · —	H · · · ·	O · ·	V · · · —
B — · · ·	I · ·	P · · · · ·	W · — —
C · · ·	J — · — ·	Q · · — ·	X · — · ·
D — · ·	K — · —	R · · ·	Y · · · ·
E ·	L —	S · · ·	Z · · · · ·
F · — ·	M — —	T —	& · · · ·
G — — ·	N — ·	U · · —	

NUMERALS.

1 · — — ·   3 · · · — ·   5 — — —   7 — — · ·   9 — · · —
2 · · — — · ·   4 · · · · —   .6 · · · · · ·   8 — · · · ·   0 ——

PUNCTUATION.

Period · · — — · ·   Interrogation — · · — ·   Paragraph — — — —
Comma · — · —   Exclamation — — — ·   Italics — · · · —
Semicolon · — · — ·   Parenthesis · — · · —

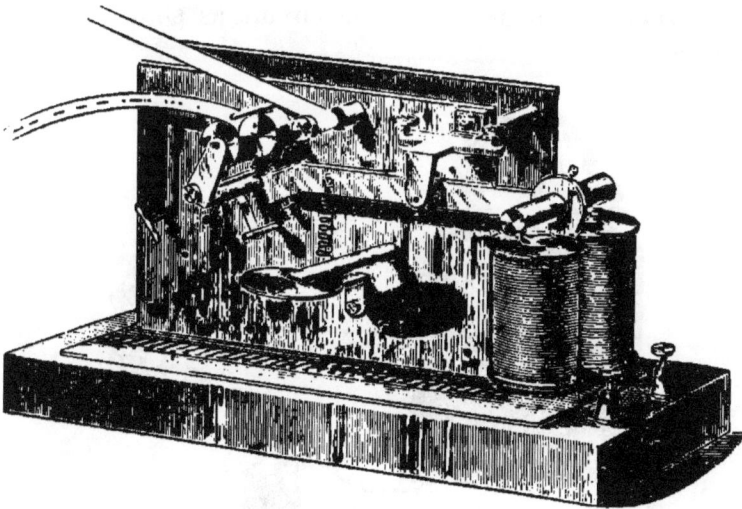

FIG. 243.

Instead of taking by sound it was formerly customary
to read the dots and dashes on a paper strip, as they were
made by a point attached to the beam bearing the armature, while the strip was kept moving by clock-work.
Fig. 243 shows the instrument which was used, in the
place of a sounder, for this purpose.

2. Electric Bells.

Construction.

Electric bells are of various kinds. Fig. 244 shows
the construction of one of the most common forms.

It consists of an electro-magnet E, in front of the poles
of which is supported an armature A by a spring S. At
the end of the armature is attached a hammer H, placed
in such a position that it will strike a bell B when the
armature is drawn in to the poles of the magnet. A
current breaker, consisting of a platinum-tipped screw C
in contact with a platinum-tipped spring D attached to
the armature, is placed in the circuit as shown in the
figure.

FIG. 244.

Action.

When the circuit is completed by a push button P,
the current from the battery passes from the screw C
to spring D, through the electro-magnet and back to
the battery. The armature is drawn in and the bell
struck by the hammer; but by the movement of the

armature the spring D is separated from the screw C, and the circuit is broken at this point. The magnet then releases the armature, the spring S causes the hammer to fall back into its original position, the circuit is again completed, and the action goes on as before. A continuous ringing is thus kept up.

3. Measurement of Current Strength — Galvanometers.

Since the magnetic effect of the current varies with its strength, the strengths of different currents may be compared by comparing their magnetic actions. Instruments for this purpose are called **galvanometers**. There are many forms of these instruments, but the following are among the most common.

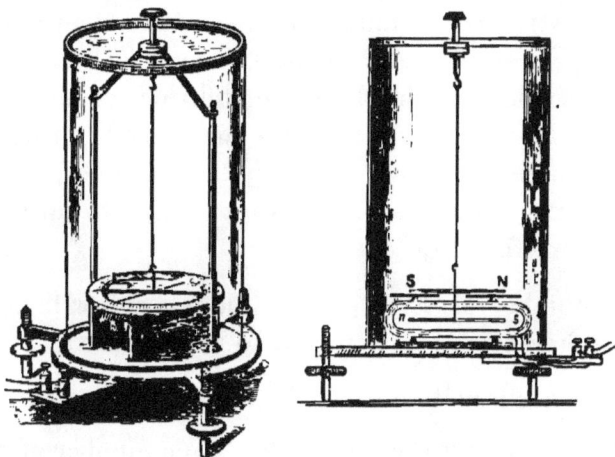

Fig. 245.

1. The Astatic Galvanometer.
Construction.

Fig. 245 shows the construction of this instrument. It consists of an astatic pair of magnetic needles, that is, a pair of magnetic needles rigidly connected as shown in Fig. 181, suspended by a silk fibre in such a position that one of the needles, usually the lower one, is free to turn around within a coil of wire, the ends of which are attached to the binding posts of the instrument. The deflection is read from a circular graduated scale placed above the coil.

Action.

When the needles are parallel with the strands of the coil, and a current passed through it, both needles will be urged in the same direction. If the deflection of the needle is not more than 15° or 20°, the strength of the current will be approximately proportional to the angle of deflection.

This instrument is much more sensitive than the galvanoscope described on page 307; because (1) the needles are independent of the directive influence of the earth's magnetism and consequently are more readily turned in the magnetic field produced by the current, (2) the current acts upon both needles and tends to turn them in the same direction.

The number of turns of wire in the coil will depend on the character of the current to be measured.

It is made sensitive to weak currents

(1) By winding the coil with a large number of turns of wire.

(2) By suspending the needles by a fibre of which the torsion is very low.

(3) By magnetizing the needles strongly.

Apply the law stated on page 348 to explain why the current in the coil tends to turn both needles of the astatic pair in the same direction.

2. The Tangent Galvanometer.

Construction.

Fig. 246 shows the construction of a tangent galvanometer. It consists of a short magnetic needle, not exceeding one inch in length, suspended, or poised, at the centre

FIG. 246.

of an open ring of copper or circular coil of copper wire not less than 12 inches in diameter. Where a large range of currents is to be measured, both the open ring and coils of different numbers of turns of wire are sometimes mounted on the same instrument as shown in the figure. A light aluminium pointer is attached to the needle, and its deflection is read on a circular graduated scale placed under the pointer.

Action.

Since the coil is large and the needle short, the magnetic field which a current passing through the ring produces is practically uniform at all points immediately surrounding the needle, and the lines of force there are at right angles to the plane of the coil (Fig. 237). On these conditions, when the coil is placed parallel with the earth's magnetic meridian and a current passed through it, **the intensity of the current will vary as the tangent of the angle of deflection of the needle.** This may be demonstrated as follows :—

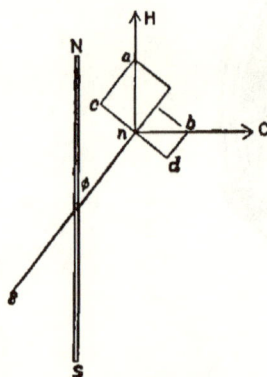

Fig. 247.

Let NS represent the coil of the galvanometer (Fig. 247);

and let ns denote the direction of the needle and ϕ the angle of deflection when a current C passes through the coil.

Two forces act upon the needle.

(i) The horizontal component of the earth's magnetism (H), acting in the direction of the magnetic meridian.

(ii) The force due to the magnetic effect of the current, acting at right angles to (i). It is proportional to the intensity of the current, and may, therefore, be represented by C.

Let the lines na and nb represent these forces in magnitude and in direction.

If the forces are resolved along two lines, one in the direction of the needle and the other at right angles to it, only the resolved parts in a direction at right angles to the needle tend to turn it around.

These resolved parts are represented by the lines nc and nd; and, since the needle is at rest

$$nc = nd,$$

but
$$nc = na \sin nac$$

$$= na \sin \phi$$

and
$$nd = nb \cos bnd$$

$$= nb \cos \phi,$$

therefore
$$na \sin \phi = nb \cos \phi,$$

or
$$\frac{nb}{na} = \frac{\sin \phi}{\cos \phi} = \tan \phi$$

or
$$nb = na \tan \phi.$$

But nb represents C, and na represents H,

therefore
$$C = H \tan \phi,$$

but H is constant.

Hence
$$C \text{ varies as } \tan \phi.$$

That is, **the intensity of the current varies as the tangent of the angle of deflection of the needle.**

If the current corresponding to any angle of deflection is known, the current corresponding to any other angle of deflection can be determined by referring to a table for the tangent of the angle, and making the necessary calculations.

Experiment.

Place in a circuit with a constant battery a tangent galvanometer and a copper voltameter, observe the reading of the
24

galvanometer and determine, as described in Art. 14, page
343, the current in amperes passing through the coil of the
galvanometer.

Make a record of the result and keep it for future experiments.

QUESTIONS.

1. If you were given a voltaic cell, wire with an insulating
covering, and a bar of soft iron, one end of which was marked,
state exactly what arrangements you would make in order to
magnetize the iron so that the marked end might be a north-
seeking pole. Give a diagram.

2. A current is flowing through a rigid copper rod. How would
you place a small piece of iron wire with respect to it, so that the
iron may be magnetized in the direction of its length? Assuming
the direction of the current, state which end of the iron will be a
north pole.

3. A strong electric current flows through a copper wire, which
passes through the centre of an iron ring, and is at right angles to
the plane of the ring. Describe the magnetic state of the ring.

4. A telegraph wire runs north and south along the magnetic
meridian. A magnetic needle free to turn in all directions is
placed beside the wire, and on the same level with it. How will
this needle act when a current is sent through the wire from south
to north? Supposing the wire to run east and west, how would
you detect the direction of a current passing through it?

5. A guttapercha covered copper wire is wound round a wooden
cylinder, AB, from A to B. How would you wind it back from B
to A, (1) so as to increase, (2) so as to diminish the magnetic effects
which it produces when a current is passed through it? Illustrate
your answer by a diagram drawn on the assumption that you are
looking at the end B.

6. An insulated copper wire is wound round a glass tube, AB,
from end to end, and a current is sent through it, which to an

observer looking at the end A, appears to go round in the same direction as the hands of a watch. A rod of soft iron is held (1) inside the tube ; (2) outside but parallel to the tube. What will be the magnetic pole at that end of the bar which is nearest to the observer in each case ?

7. Two parallel covered wires are traversed by equal currents in the same direction : what is the joint effect of the currents upon a bar of soft iron (a) laid across the two wires, on the same side of both ; (b) held between the wires at the same distance from each ?

8. Two compass needles are arranged near each other so that both point along the same straight line. A wire connecting the platinum and zinc ends of a battery is stretched vertically half-way between the needles. How will the current in the wire affect the needles, and how will the result depend upon whether the platinum terminal is connected with the upper or the lower end of the wire?

9. Two long wires are placed parallel to each other in the same horizontal plane, and in the magnetic meridian. A magnetic needle, capable of turning in any direction about its point of suspension, is placed exactly half-way between them. How will it behave, if the same electric current flows through the easterly wire from south to north, and through the westerly wire from north to south? (The action of the earth on the magnetic needle may be neglected.)

10. One end of a coil of wire, through which a current passes, is found to attract the north pole of a compass-needle, when placed at a certain distance from it. Will the action be the same (1) in nature, (2) in amount, when a rod of soft unmagnetized iron is placed inside the coil ?

11. A number of galvanic cells are connected together in a row to form a battery. This row is laid on a table so as to lie north and south. The zinc is to the north. The poles of the battery are connected together by a wire, which passes from one pole, up one wall of the room, across the ceiling, and down the opposite wall to the other pole of the battery. How will a magnetic needle be affected which is placed under the table and just below the battery ?

12. A coil of wire is suspended in front of the one pole of a bar magnet. A current is made to flow in the coil. How will the coil move (1) when its axis points in the line of the magnet, (2) when it points at right angles to that line ?

13. A small coil is suspended between the poles of the powerful electro-magnet and is movable about a vertical axis. How will it move when a current flows in it ? If set in a certain position it will not move. What is this position ?

14. If it were true that the earth's magnetism is due to currents traversing the earth's surface, show what would be their general direction.

15. An elastic spiral of wire hangs so that its lower end just dips into a vessel of mercury. Describe and explain what happens when the top of the spiral is connected with the one pole and the mercury is connected with the other pole of a battery.

CHAPTER XXVIII.

I.—Induced Currents.

1. Apparatus.

Prepare two large coils of the form shown in Fig. 248, by winding double cotton covered magnet wire No. 16 on wooden spools of which the dimensions are, length 4 inches, diameter of flanges 3 inches, diameter of the opening through the spool

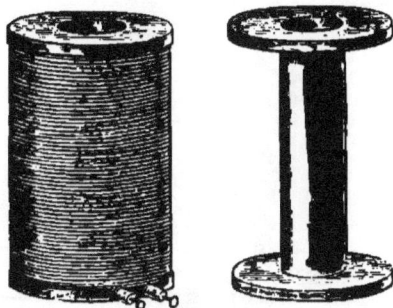

FIG. 248.

1 inch, thickness of walls $\frac{1}{16}$ inch, thickness of flanges $\frac{1}{2}$ inch. The wire should be carefully wound on each in the same direction, and the ends attached to binding posts, as shown in the figure.

Also prepare another coil (Fig. 249), made by winding magnet wire No. 35 on a spool of the following dimensions: length 4 inches, diameter of flanges $1\frac{5}{8}$ inch, diameter of opening through the spool $\frac{7}{8}$ inch, with the thickness of walls and flanges as in the large spools.

Obtain two soft iron rods, one 10 inches long, which will just pass easily through the opening in

FIG. 249.

[373]

the large spool, and another 5 inches long, which will just pass
through the opening in the small spool.   Also obtain two soft •
iron rods $4\frac{1}{4}$ inches long and 1 inch in diameter, to be connected as shown in Fig. 250 by an iron plate, 6 inches long
and $\frac{1}{2}$ inch thick, the centres of the rods being $4\frac{1}{2}$ inches apart,
and their upper ends being bored and tapped to receive bolts.
The plate should be secured to a wooden stand and when the
large coils are fitted over the rods the apparatus should appear
as shown in Fig. 253.

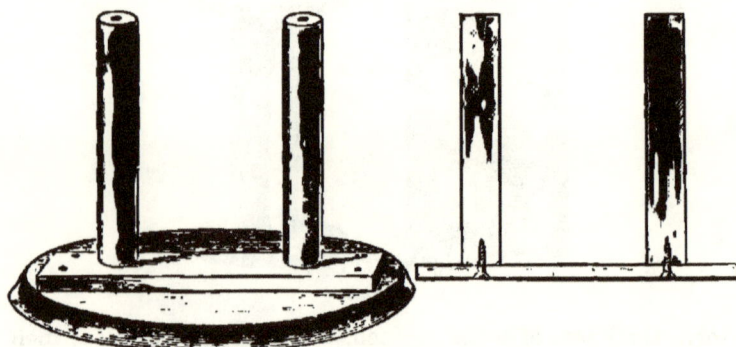

FIG. 250.

2. Production of Induced Currents.

Experiment 1.

Connect the ends of the wire of the large coil with a sensitive galvanometer and thrust (1) slowly, (2) quickly, a powerful bar magnet into the opening of the spool (Fig. 251).

1. Does the galvanometer indicate a current (1) at the instant the
magnet is thrust into the coil, (2) while it remains stationary in the
coil?

Withdraw the magnet (1) quickly, (2) slowly.

2. Does the galvanometer indicate a current at the instant the
magnet is withdrawn?

3. If a current is produced by (a) inserting, and (b) withdrawing
the magnet from the coil, does it flow in the same direction in each

case, and is there any relation between the E.M.F. of these currents and the rapidity with which the magnet is inserted or withdrawn ?

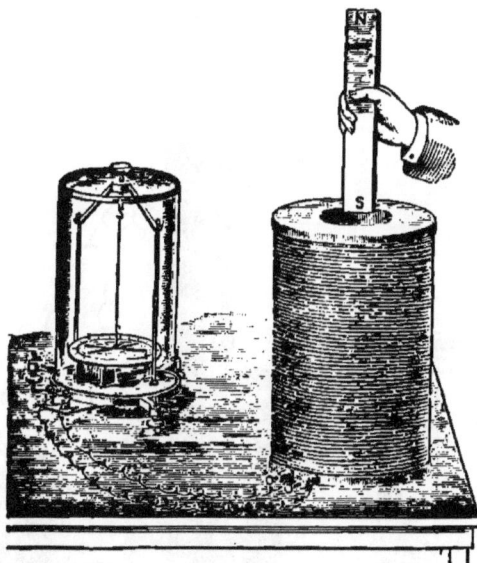

FIG. 251.

4. Does the inserting of the magnet increase or diminish the number of magnetic lines of force which pass across the space enclosed by the coil ? What effect has withdrawing it upon the number of lines of force passing across this space ?

Experiment 2.

Make an electro-magnet by placing the soft iron rod within the small coil of fine wire prepared as described above and connecting the ends of the wire with a battery. Place the galvanometer in the circuit for an instant and observe the direction of the deflection of the needle. Now remove the galvanometer from this circuit and connect it with the large coil, as shown in Fig. 252, taking care to connect each binding post with the end of the wire of the large coil which corresponds to the end of the small coil with which it was connected.

When the small coil is connected with the battery, thrust it quickly into the large coil, allow it to stand a few seconds and then withdraw it quickly. Repeat the experiment several times, changing the rapidity with which the coil is inserted and withdrawn.

Fig. 252.

1. When does the galvanometer indicate that a current is flowing through the coil connected with it?

2. When does this current flow in the same direction as the battery current, when the coils are approaching or when they are receding from each other?

3. What is the relation between the rapidity with which the coils are brought together, or separated, and the E. M. F. of the current produced in the coil connected with the galvanometer?

3. Explanation of Terms.

The coil connected with the battery is called the **primary coil**, and the current which flows through it is called the **primary current**; the coil connected with the galvanometer is called the **secondary coil**, and the momentary currents made to flow in it, **secondary currents**. When the secondary currents flow in the same direction as the primary, they are said to be "**direct**," or to flow in a **positive direction**; but when the secondary currents flow in the opposite direction, they are said to be "**inverse**," or to flow in a **negative direction**.

Is the secondary current direct or inverse when the number of lines of force passing through the space enclosed by the secondary coil is (1) increasing, (2) decreasing ?

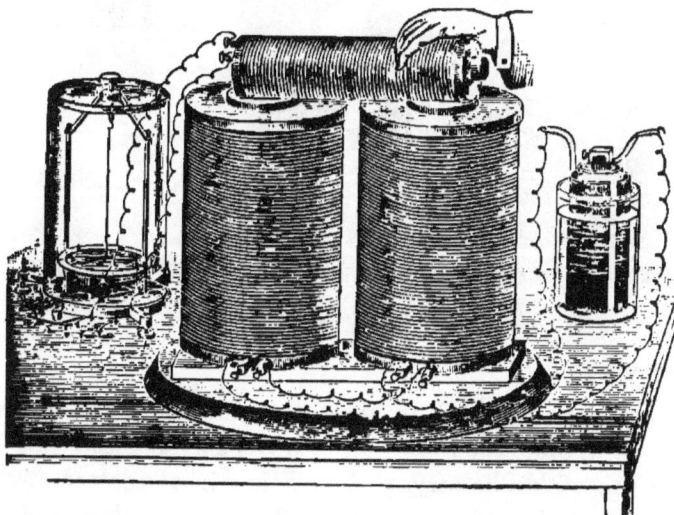

FIG. 253.

Experiment 3.

Repeat Experiment 2, connecting the battery with the outer coil and the galvanometer with the inner coil.

Experiment 4.

Place the two large coils on the vertical rods shown in Fig. 250, and connect them as shown in Fig. 253, thus forming a large electro-magnet with opposite poles at the upper ends of the rods. Connect with the battery as shown in the figure. Place the iron rod within the small coil and connect the terminals with the galvanometer. Hold this coil over the poles of the electro-magnet, and, keeping its axis in line with the poles of the magnet, move it backward and forward and turn it end for end.

On what conditions are (1) direct, (2) inverse currents produced in the coil?

Fig. 254.

Experiment 5.

Place .the small coil within the large one, insert the iron rod, and connect the small coil with the galvanometer and the large one with a battery, placing a key in the latter circuit, as shown in Fig. 254. Quickly make and break the circuit two or three times with the key.

1. What kind of current is produced in the secondary coil (1) when the circuit is completed, (2) when it is broken?

2. How does (1) completing the circuit, (2) breaking it, change the number of lines of force passing through the space enclosed by the coil?

3. In which case, therefore, is a direct secondary current produced in the secondary coil, when the number of lines of force passing through the space enclosed by the current is increasing or diminishing?

Experiment 6.

Repeat the last experiment, connecting the outer coil with the galvanometer and the inner one with the battery.

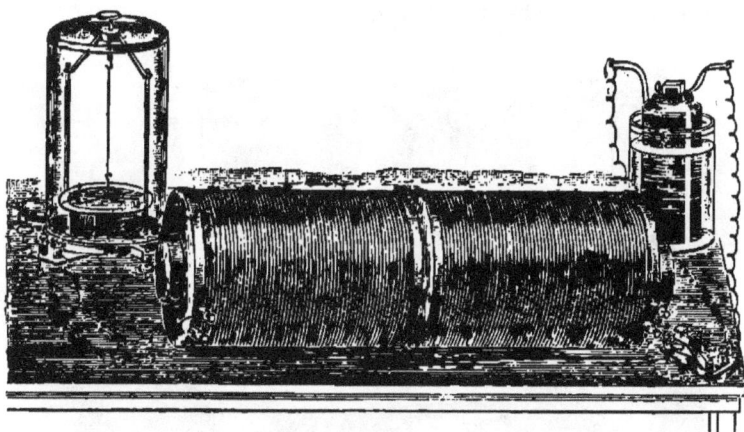

FIG. 255.

Experiment 7.

Repeat Experiments 5 and 6, placing in the circuit a rheostat instead of the key.

Alternately lessen and increase the current given by the battery by increasing and decreasing with the rheostat the resistance in the circuit.

1. Does any change in the strength of the current passing through the coil connected with the battery cause a current in the other coil?

2. If so, when is a direct current produced, and when an inverse?

Experiment 8.

Repeat Experiments 5, 6 and 7, using the two large coils instead of the large one and small one, and (1) placing them end to end, as shown in Fig. 255, with an iron rod through the openings, (2) placing them on the upright rods, as shown in Fig. 256.

Experiment 9.

Place the coils again as in Experiment 5. Connect the outer one with the battery and the inner one with the galvanometer, and move backward and forward in front of the iron rod a plate of soft iron.

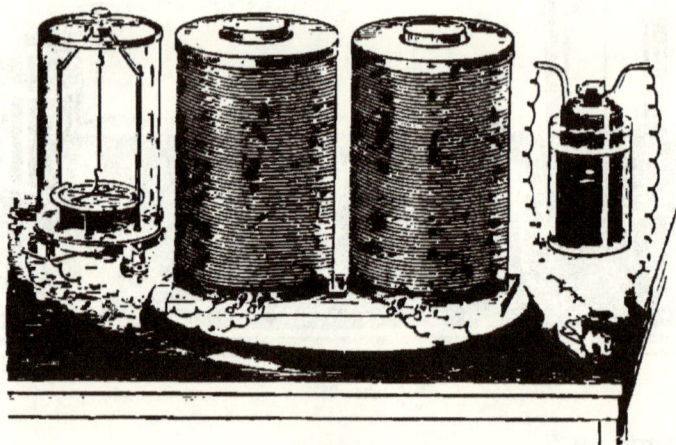

FIG. 256.

1. Does the movement of the iron plate cause a current in the inner coil?

2. How does the movement of the iron plate cause a change in the magnetic field surrounding the inner coil?

4. Summary.

These experiments show that

1. **Any change in the number of lines of forces pass-**

ing through the spaces enclosed by a coil of wire produces a current of electricity in the wire.

When a current is set up in a coil of wire in this way, it is said to be produced by **induction.**

The following are the laws:—

Laws of Induction.

1. Whenever a decrease in the number of lines of force which pass through a closed circuit takes place, a current is induced in this circuit flowing in the same direction as that which would be required to produce this magnetic field, that is, a direct current is produced.

Whenever an increase in the number of lines of force takes place, the current induced is such as would by itself produce a field opposite in direction to that acting; that is, an inverse current is produced.

2. The total electromotive-force induced in any circuit at a given instant is equal to the time-rate of the variation of the flow of magnetic lines of force across this circuit.

5. Self-Induction.

Experiment.

Place the large coils arranged as an electro-magnet, as shown in Fig. 253, in a circuit with a battery. Close and open the circuit by holding the ends of the wires in your hands, touching them together, and then separating them.

1. What do you observe at the ends of the wires when they are separated?

2. Is a shock felt? Dampen your fingers and repeat the experiment, grasping the bare wires.

The effects observed are due to what is known as **self-induction.**

The magnetic lines of force surrounding a current in circulating around the wire pass, especially when the wire is coiled, across contiguous parts of the same circuit, and any variation in the strength of the current causes the current to act inductively on itself. On completing the circuit, this current is inverse; and on breaking it, direct.

The direct induced current in the primary wire itself, which tends to strengthen the current when the circuit is broken, is called the **extra current.**

This self-induced current is of high E.M.F., and therefore flows for an instant across the air space when the wires are a short distance apart; hence the spark.

1. Why will an iron core placed within a coil of wire in a circuit increase the intensity of the extra current when the circuit is broken?

2. If a large electro-magnet is placed in a circuit with a galvanometer and a secondary battery, on closing the circuit the current will be seen to rise gradually and take it full strength only after several seconds. Explain.

3. If you place across the terminals of a large electro-magnet an incandescent lamp of such resistance that a battery current will bring it only to dull redness, and you break the connection with the battery, you will observe that the lamp will become vividly incandescent for an instant. Explain. Try the experiment.

QUESTIONS.

1. You have a metal hoop. Describe (and give a figure of) some arrangement by which, without touching the hoop, you could make electric currents pass round it, first one way and then the other.

2. A bar of perfectly soft iron is thrust into the interior of a coil of wire whose terminals are connected through a galvanometer. An induced current is observed. Could the coil and bar be placed

in such a position that the above action might nearly or entirely disappear? Explain fully.

3. A piece of covered wire is passed a few times round a wooden hoop; its ends are joined up to a galvanometer. The ends of another piece of covered wire which is wrapped round a similar hoop are joined up to a battery. What will happen if the two hoops are (1) brought quickly near to each other, and (2) if they are quickly separated?

4. How could you temporarily stop or weaken a current in a wire without disconnecting it from the battery, by means of the motion of another wire through which a current is passing?

5. The poles of a voltaic battery are connected with two mercury-cups. These cups are connected successively by

(1) A long straight wire;

(2) The same wire arranged in a close spiral, the wire being covered with some insulating material;

(3) The same wire coiled round a soft iron core.

Describe and discuss what happens in each case when the circuit is broken.

6. Round the outside of a deep cylindrical jar are coiled two separate pieces of fine silk-covered wire, each consisting of many turns. The ends of one coil are fastened to a battery, those of the other to a sensitive galvanometer. When an iron bar is thrust into the jar a momentary current is observed in the galvanometer coils, and when it is drawn out another momentary current, but in an opposite direction, is observed. Explain these observations.

7. A small battery was joined in circuit with a coil of fine wire and a galvanometer, in which the current was found to produce a steady but small deflection. An unmagnetized iron bar was now plunged into the hollow of the coil and then withdrawn. The galvanometer needle was observed to recede momentarily from its first position, then to return and to swing beyond it with a wider arc than before, and finally to settle down to its original deflection. Explain these actions, and state what was the source of the energy that moved the needle.

II.—Practical Applications.

1. Ruhmkorff Induction Coil.

The Ruhmkorff induction coil is an instrument by means of which currents of high E.M.F. are produced by

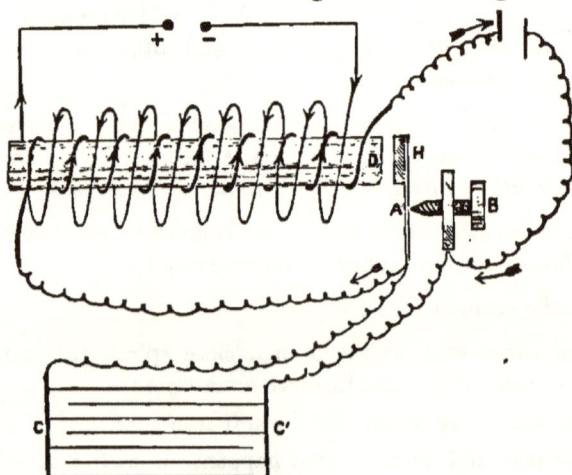

Fig. 257.

the action of an electric current in a circuit which is alternately opened and closed in rapid succession.

Fig. 257.

Construction.

Fig. 257 shows the essential parts of the instrument. A primary coil, consisting of a few turns of stout insulated copper wire, is wound around a core D made up of a bundle of soft iron wires. One end of this coil is attached directly to one of the poles of a battery ; and the other end is connected to the other pole of the battery by means of a current breaker, which consists of a hammer H supported in front of the iron core by a spring A in contact with a screw B, the wires being connected as shown in the figure. The spring and screw are also connected with a condenser C, C¹ made of alternate layers of tinfoil and paraffined paper, in such a manner that one is joined to the even sheets of foil and the other to the odd ones.

A secondary coil, consisting of a great number of turns of very fine insulated wire, is wound on the outside of the primary coil. The terminals of this coil are attached to binding posts placed above the coil.

Action.

When the primary circuit is completed and the battery current passes through the coil, the core is magnetized, the hammer is drawn in, and the circuit broken between the spring A and the screw B. The hammer then falls back, the circuit is completed, and the action goes on as before. An interrupted current is thus sent through the primary coil, which induces currents of high electro-motive-force in the secondary coil.

The function of the condenser is to prevent the extra current induced in the primary coil, when the circuit is broken, from passing across the break in the form of a spark, and prolonging the time of fall to zero of the primary current, in consequence of which the rate of variation of the flow of the magnetic lines of force

25

across the secondary coil would be diminished and the electromotive-force of the induced current lowered. The extra current flows by the shunt circuit into the condenser, and causes a difference in potential between the layers of foil connected with the two wires. This immediately gives rise to a current in the opposite direction which flows back through the primary coil and instantaneously demagnetizes the soft iron core, thus causing the inverse current induced in the secondary coil, when the primary circuit is broken, to become shorter and more intense.

The potential-difference between the terminals of the secondary coil can in this way be made sufficiently great to cause a spark to pass between them when they are placed a short distance apart.

This potential-difference is increased by increasing the number of turns of wire, by increasing the current in the primary coil, and by decreasing the time required for it to fall to zero each time the circuit is broken.

The iron core, on account of its magnetic permeability, increases the number of lines of force passing through the coils.

A bundle of iron wire is used instead of a solid bar of iron to produce a stronger magnetization, and to prevent the circulation of induced currents in the iron itself.

2. The Transformer.

When the primary current is of constantly varying strength, or is an alternating current, the current breaker in the primary circuit is unnecessary, and the apparatus for transforming a current of one electromotive-force to that of another consists of but the two coils and the iron core, as shown in Fig. 258. There are many forms of this instrument, but the essential parts of all are the same—two coils and a laminated soft iron core, so placed that as many as possible of the

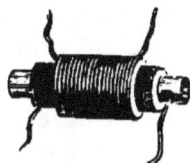

Fig. 258.

lines of force produced by the current in one coil will pass through the space enclosed by the other.

When the current is alternating, the electromotive-forces of the currents generated in the secondary coil are to those of the primary currents nearly in the ratio of the number of turns of wire in the secondary coil to the number in the primary.

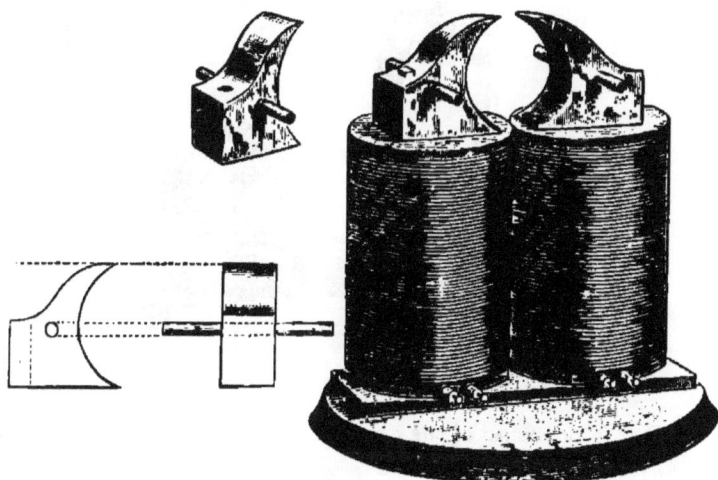

Fig. 259.

3. The Dynamo.

Experiment 1.

Provide two cast iron pole-pieces of the form shown in Fig. 259, to be attached by bolts to the large electro-magnet

used in former experiments.   The circular opening between the
pole-pieces should be about $3\frac{1}{2}$ inches in diameter.   Provide
also a soft iron ring, about $2\frac{3}{4}$ inches in diameter and of the
form shown in Fig. 260.   It is best made of soft iron wire,
but a ring of solid iron will answer.   Insulate the ring by
covering it with the insulating tape used for wrapping joints,
and wind a coil of a number of turns of fine insulated copper
wire around it, as shown in the figure.   Connect the ends of
the wire with a sensitive galvanometer by means of flexible
conducting cords, and insert the coil into the circular opening
between the pole-pieces.   Connect the electro-magnet with a
battery, as shown in figure, revolve the ring, and observe the
directions of the current generated in the coil.   Repeat the
experiment several times.

Fig. 260.

1. What is the cause of the currents produced in the coil ?   (See
Art. 4, page 381.)

2. When the coil is being turned around, at what positions (1)
is the direction of the current in it changed, (2) is the E.M.F.
of the induced current at a maximum ?

3. Does an increase in the velocity of revolution increase the
E.M.F. of the induced current ?   If so, why ?

Experiment 2.

Obtain another ring similar to that used in the last experiment, and, after insulating it as before, wind upon it a number of coils of insulated copper magnet wire No. 20, as shown in Fig. 261. Fit the ring over a wooden hub, on one

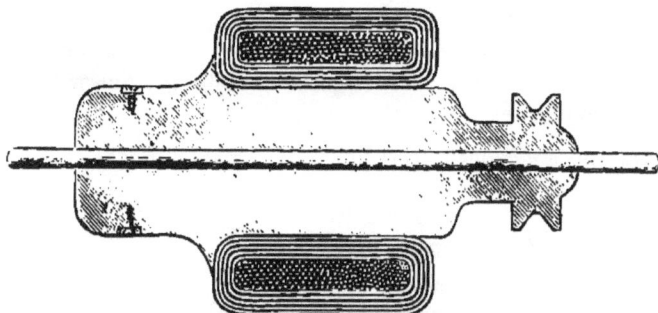

Fig. 261.

end of which is turned a pulley, and through the centre of which passes an iron or brass shaft. Connect the end of each coil with the beginning of the next by fastening the wires to small copper plates screwed to the hub, as shown in the figure. The plates are to be placed side by side, but must not be allowed to touch. Mount the axle on cross bars screwed to the pole-pieces of the large electro-magnet, and attach copper strips, to serve as brushes, to binding posts held in position by a vulcanite bar (Fig. 262). Turn the

Fig. 261.

brushes so that they will just touch the small copper plates, and connect them with a galvanometer. Now connect the electro-magnet with a battery, and, by means of a whirling-machine, revolve the ring of coils.

1. What evidence have you that the current which passes from one brush to the other through the galvanometer flows always in the same direction?

2. Why does this current flow in one direction?

To answer this question, consider the following results of the previous experiment :—

1. The currents in the single coil are always flowing in the same direction when it is on the same side of a line drawn through the centre of the ring nearly at right angles to the line joining the poles of the magnet.

2. On opposite sides of this line the currents in the coil are in opposite directions.

Fig. 262.

3. What, therefore, may be inferred with regard to the directions in which the currents tend to flow (1) in each coil on the same side of this line, (2) in the connected coils on opposite sides of the line?

4. If the brushes are so placed that one touches the wires at the point from which and the other at the point to which these currents tend to flow, in what direction will the current flow in the wire which joins the brushes?

5. Where should the brushes be placed ? Why ?

6. Why does the current flow in the same direction in all the coils on the same side of a line drawn nearly at right angles to the line joining the poles of the magnet, and in the opposite direction on the opposite side of the line ? (See Art. 4, page 381.)

Experiment 3.

Repeat the last experiment, disconnecting the magnet from the battery and connecting it with wires from the brushes, as shown in Fig. 262.

1. Is a current produced when the ring of coils is revolved ?

2. If so, how is the magnet excited ?

6. The Direct Current Dynamo.

The preceding experiments illustrate the method of generating an electric current by a dynamo.

Construction.

Fig. 263 shows the essential parts of a direct current dynamo. It consists of a **field-magnet** between the poles N, S of which is revolved an **armature,** consisting of a laminated soft iron core A, around which are wound coils c, c, c . . . of insulated copper wire. The ends of these coils are joined to copper **commutator plates** P, P, P . . . as shown in the figure. These plates are insulated from one another and are rubbed at points about midway between the poles by copper or carbon **commutator brushes B B₁.**

The wires of the main circuit are connected with the brushes, and the field-magnet coils are either connected in **series,** that is, made a part of the main circuit as shown in Fig. 264, or are in a **shunt circuit,** as shown in Fig. 265.

Action.

When the circuit is closed and the armature revolved, currents will be induced in the connected coils by the constant variations in the number of lines of force pass-

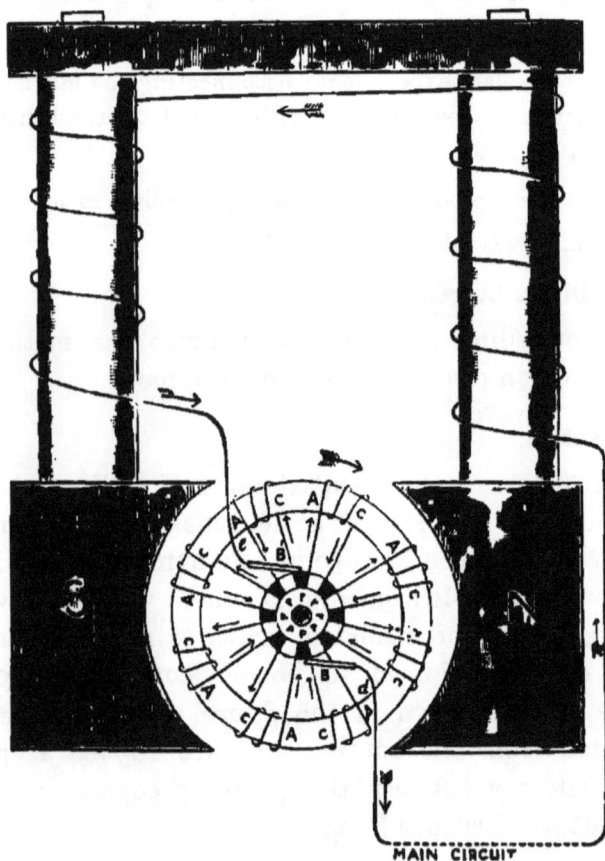

Fig. 263.

ing through the space enclosed by each coil. The direc- tions of these currents are indicated by the arrows in Fig. 263. A continuous current, therefore, flows from the brush which presses the commutator plate at B.

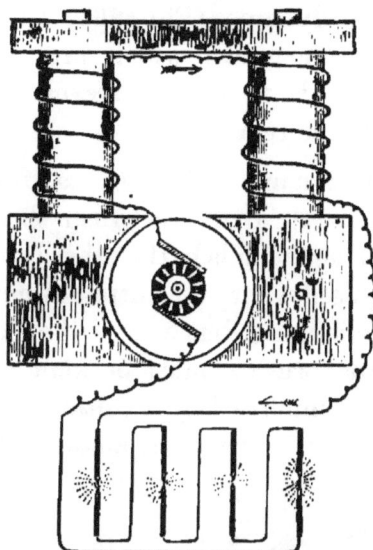

FIG. 264.

through the main circuit to the brush which presses the plates at B_1.

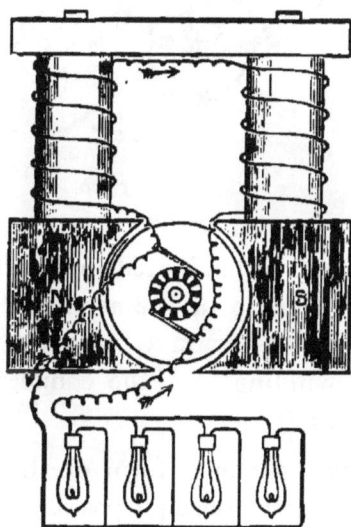

FIG. 265.

In the series dynamo (Fig. 264) the full current excites the field magnets, which are usually wound with coarse wire. It is used where, as in arc lighting, a constant current is required.

In the shunt-dynamo (Fig. 265) a fraction of the current passes directly from one brush through the high resistance field-magnet coils to the other brush. Dynamos of this class are used where the output of current required is continually changing, but where the potential-difference between the brushes must be kept constant. The regulation is accomplished by a suitable rheostat placed in the shunt circuit to vary the amount of the exciting current.

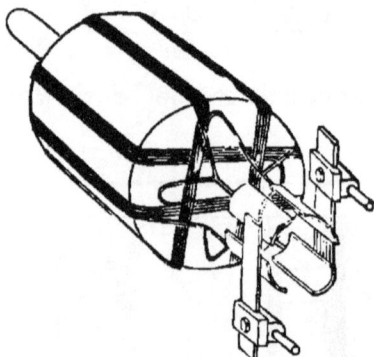

FIG. 266.

A cylindrical core of laminated iron is frequently used instead of the ring core in the armature of a dynamo. It is then known as the **drum armature**. Fig. 266 shows the method of winding and the connections with the commutator plates.

Fig. 267 shows a common form of the direct current dynamo used for commercial purposes.

QUESTIONS.

1. Upon what is the potential-difference between the brushes of a dynamo dependent?

Fig. 267.

2. To what is the internal resistance of a dynamo due?

3. How should a dynamo for producing currents at a high electro-motive-force be wound?

4. How should a dynamo used to produce a current for electro-plating be wound?

5. What would be the effect of short-circuiting (1) a series-dynamo, (2) a shunt-dynamo? Explain.

6. What would be the effect upon the potential-difference between the brushes of a dynamo of moving them backward or forward around the ring of commutator plates? Explain.

7. Why is the armature core laminated?

8. What would be the effect upon the working of a dynamo of connecting the commutator plates by binding a bare copper wire around them, (1) if the field-magnet coils are in a shunt circuit, (2) if the field-magnet is excited by a separate dynamo? Would a current be generated in either of these cases? If so, where would it flow?

---

4. The Electric Motor.

The direct current dynamo may be used as a **motor**. The armature and field-magnet coils are wound to suit the electromotive-force of the current used.

Action of the Dynamo as an Electric Motor.

The current supplied to the motor divides at d, Fig. 268; part flows through the field-magnet coils and part enters the armature coils by the brush B at the point b, where it divides, part passing through the coils on one side of the ring, and part through the coils on the other side. The currents through the armature coils re-unite at a, pass out by the brush B_1, and are joined at c by the part of the main current which flows through the field-magnet coils.

Both the field-magnet and the armature cores are thus magnetized, and poles are formed according to the law stated in Art. 4, page 351. The poles of the field-magnet are as indicated in the figure. Each half of the iron core will be an electro-magnet of the horse-shoe type, each having a south pole at a and a north pole at b.

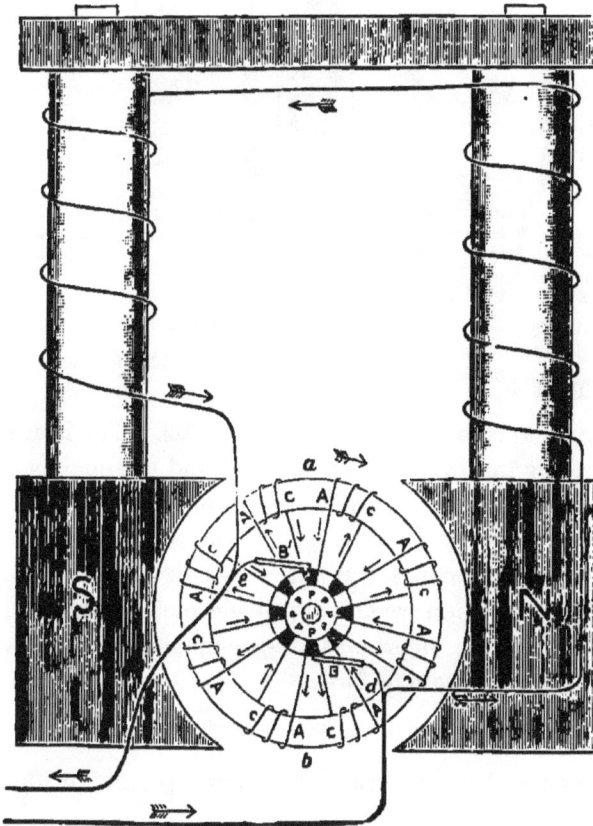

Fig. 268.

The mutual attractions and repulsions between the poles of the armature and the field-magnet cause the armature to revolve.

Trace as far as you can the transferences and transformations of energy in the following :—

A printing press is driven by an electric motor to which the current is supplied by a dynamo driven by a steam engine.

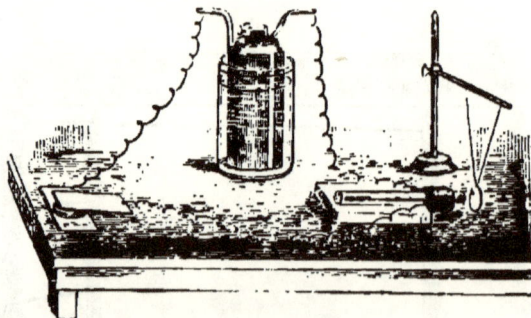

Fig. 269.

5. The Telephone.

Experiment 4.

Arrange apparatus as shown in Fig. 269. The iron rod has a coil of fine insulated wire wound around one end of it, and in front of this end is suspended a small soft iron disc. A battery is placed in the circuit with the coil, and the circuit is completed by attaching the wires to two small metal plates separated by a small wedge-shaped piece of graphite, or stove polish. Place the finger on the upper plate, and press upon it with different degrees of force.

Describe and explain the movements of the disc.

Experiment 4 illustrates the principle of action of the telephone.

Construction.

Fig. 270 shows the working parts of the telephone now most commonly used.

The **transmitter** consists of a thin iron diaphragm D held against a small convex platinum button B, which is pressed by a highly polished carbon button A mounted in a piece of brass. The platinum and carbon buttons are supported by springs S, S¹.

The **receiver** consists of a magnetized steel rod near one end of which is wound a coil of fine insulated copper wire. In front of this end of the rod is supported a thin iron diaphragm D¹.

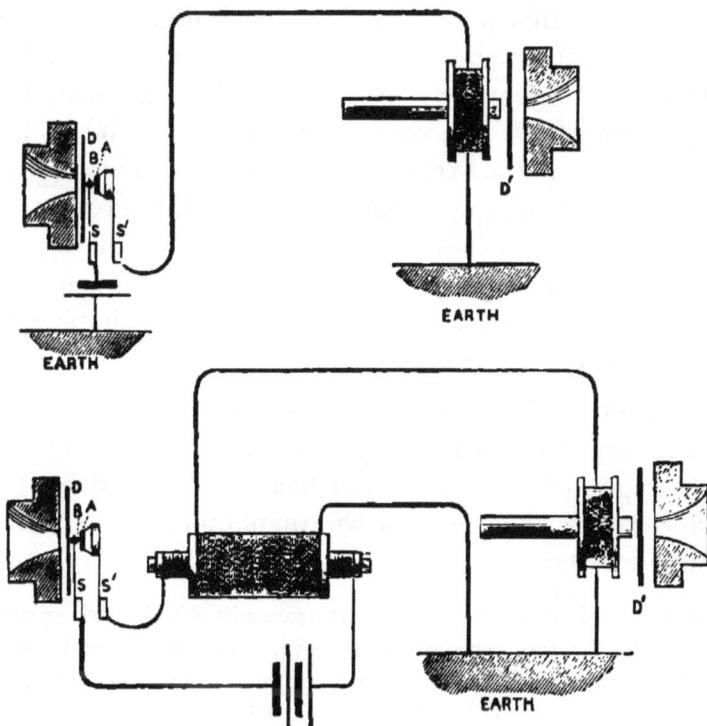

Fig. 270.

The receiver coil is connected by a wire with the carbon button, and the platinum button is connected with a battery. The earth usually forms the return current, as shown in the figure.

For long distance lines a transformer, to increase the electromotive-force of the current given by the battery, is placed in the circuit, as shown in the lower figure.

Action.

Sound-waves cause the diaphragm of the transmitter to vibrate. When it moves forward, the pressure between the platinum and carbon buttons is increased, and the resistance at this part of the circuit is decreased. The strength of the current passing through the coil of the receiver is consequently increased, and, as a result, the diaphragm of the receiver is drawn inward. When the diaphragm of the transmitter moves backward, the pressure between the buttons is decreased, the resistance is, therefore, increased, and the current in the circuit decreased. Through the decrease in current the magnet in the receiver loses some of its power, and the diaphragm in front of it springs backward.

Hence the vibrations of the diaphragm of the transmitter are accompanied by similar vibrations of the diaphragm of the receiver, which will reproduce the sound-waves which caused the diaphragm of the transmitter to vibrate.

In the original Bell telephone the transmitter and receiver were similar to the receiver described above, and the two coils were connected by wires without a battery. Explain how this transmitter generates a current and give the reasons for the vibration of the diaphragm in the receiver. (See Experiment 9, page 380.)

CHAPTER XXIX.

HEATING AND LIGHTING EFFECTS OF THE ELECTRIC CURRENT.

I.—The Electric Current and Heat.

Experiment 1.

Connect three or four cells of a battery as shown in Fig. 271. Attach a copper wire to each pole, and complete the circuit by attaching to the free end of one of the copper wires a piece of fine platinum or iron wire four or five inches long, .

Fɪɢ. 271.

and touching the end of the other copper wire to the end of the platinum or iron wire. (The fine iron wire used by florists answers well.) Slide the copper wire along the iron wire up toward the other copper wire.

What evidence have you of the production of heat ?

Experiment 2.

Find by trial the length of a fine wire that your battery will heat red-hot, and place a piece about one-half this length

26   [401]

in a circuit with the battery and a rheostat. Regulate the current with the rheostat so that the wire will again be just red-hot. Increase the current by decreasing the resistance in the circuit.

What effect has increasing the current upon the temperature of the wire ?

Experiment 3.

Place a piece of wire about one-third the length of that which your battery will heat red-hot in a circuit with the battery, a rheostat, and a tangent galvanometer. Regulate the current with the rheostat so that the wire will again be just red-hot. Now increase the length of the wire gradually, keeping the wire as nearly as possible at the same temperature by decreasing the rheostat resistance in the circuit. Observe the current strength required to heat to the same temperature each length of wire.

1. Is any additional current required to heat the increased length of wire ?

2. How will the potential-difference between the two ends of the short wire compare with that between the two ends of the longer wire when they are both heated to the same temperature ? Why ?

Whenever an electric current meets with resistance in a conductor heat results; and, as no body is a perfect conductor of electricity, a certain amount of the energy of the electric current is always transformed into the energy of molecular motion.

Joule, who has investigated this subject, found, by comparing the results of numerous experiments, that **the number of units of heat developed in a conductor varies as**

(1) **Its resistance.**

(2) **The square of the strength of the current.**

(3) **The time the current flows.**

1. Practical Applications.

Resistance wires heated by an electric current are used for a variety of purposes, such as performing surgical operations, igniting fuses, cooking, heating electric cars, etc.

Rods of metal are welded by pressing them together with sufficient force while a strong current of electricity is passed through them. Heat is developed at the point of junction, where the resistance is the greatest, and the metals are softened and become welded together.

II.—The Electric Current and Light.

There are two systems of electric lighting, the incandescent and the arc.

EXHAUSTED GLASS GLOBE

HIGH RESISTANCE CARBON FILAMENT

PLATINUM WIRES SEALED IN GLASS

INSULATING CEMENT

FIG. 272.

1. The Incandescent Lamp.

Construction.

Fig. 272 shows the construction of the incandescent lamp. A carbon filament, made by carbonizing a thread of bamboo or other fibre at a very high temperature, is attached to conducting wires and enclosed in a pear-shaped glass globe, from which the air is then exhausted. The conducting wires are of platinum where they are fused into the glass.

Action.

When a sufficient current is passed through the high resistance carbon filament, it is heated to incandescence and yields a bright, steady light. The carbon is infusible, and does not burn for lack of oxygen to unite with it.

Grouping of Lamps.

All the lamps to be used in the same circuit are so constructed as to give their proper candle-power when the same potential-difference is maintained between their terminals. This is generally from 100 to 110 volts. The lamps are connected in multiple, or parallel, that is, the current from the leading wires divides, and a part flows through each lamp, as shown in Fig. 265. The dynamo is regulated to maintain a constant potential-difference between the leading wires.

If each lamp (Fig. 265) requires one-half of an ampere of current to raise it to its proper candle-power, what current flows through each of the following wires : (1) from the brushes of the dynamo to the first lamp, (2) between the first and second lamps, (3) between the third and fourth lamps.

2. The Arc Lamp.

The Arc Light.

When two carbon rods, or pencils, are connected by conductors with the poles of a sufficiently powerful battery or dynamo, touched together, and then separated a short distance, the current continues to flow across the gap, developing intense heat and raising the terminals to incandescence, thus producing a powerful light, generally known as the **arc light.**

Explanation.

When the carbon points are separated by air only, the potential-difference between them, when connected with the poles of an ordinary arc-light dynamo, is not sufficient to cause a spark to pass, even when they are very close together; but when they are in contact, and then separated while the current is passing through them, the "extra current" spark, produced on separation, volatalizes a small quantity of the carbon between the points, and a conducting medium, consisting of carbon vapour and heated air, is thus produced, through which the current continues to flow.

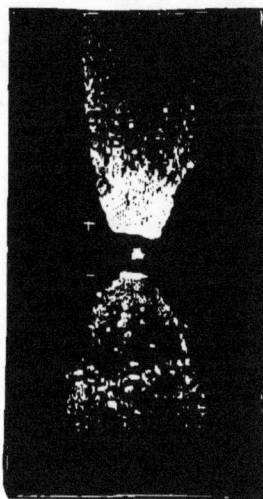

FIG. 273.

Since this medium has a high resistance, intense heat is developed and the carbon points become vividly incandescent, and burn away slowly in the air. When a direct current is used, the point of the positive carbon becomes hollowed out in the form of a crater, and the negative one becomes pointed, as shown in Fig. 273.

The greater part of the light is radiated from the carbon points, the positive one being the brighter.

Grouping, Regulation, Etc.

Arc lamps are usually connected in series, that is, the negative carbon of one lamp is connected with the positive carbon of the next, as shown in Fig. 264, and a constant current flows from one pole of the dynamo through each of the lamps to the other pole.

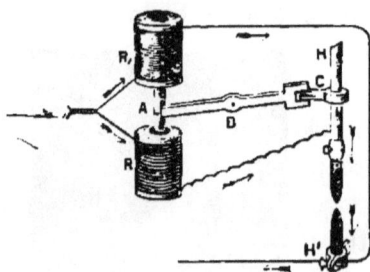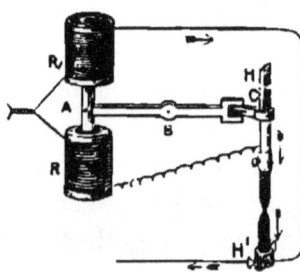

Fig. 274a.   Fig. 274.

Regulators for maintaining a constant distance between the carbon points as they burn away, are of a great variety of patterns. In most of them the regulation is accomplished by the action of two electro-magnets, or solenoids, of which the one is in series with the carbons and the other in a shunt-circuit between them. Fig. 274 shows a simple form of regulator. When the carbons are together, and the circuit closed, the current passes through the coarse wire solenoid R to the positive carbon holder H, and through the carbon to the lower carbon holder H^1 and to the conductor. The iron plunger A is drawn down and the ring-clutch C attached to the pivoted beam B catches the positive carbon holder H, lifts it, and thus separates the carbons (Fig. 274a). A very small fraction of the current passes through the fine wire high resistance solenoid R_1 directly to the lower holder. As the carbons burn away and the resistance of the arc increases, sufficient current is shunted through this fine wire solenoid to draw the plunger up and thus to release the grip of the clutch on the holder, which then falls by its own weight and shortens the length of the arc. The distance between the points is thus kept fairly constant.

A potential-difference between the carbons of about 50 volts is required to maintain an arc light, and a current of from 9 to 10 amperes is usually employed.

If a current of 10 amperes passes through each lamp (Fig. 264), where there are four lamps in series, what current passes from the one brush of the dynamo to the other ?

CHAPTER XXX.

I.—Ohm's Law.

We have learned (Art. 14, page 311) that the **strength of a current**, or the **quantity of electricity which flows past a point in a circuit in one second**, is dependent on the E. M. F. of the current and the resistance of the circuit. The exact relation among these quantities was first enunciated by Ohm. It may be thus stated:

Ohm's Law.

The current varies directly as the electromotive-force and inversely as the resistance of the circuit.

Practical Unit.

The **unit resistance** is the **ohm**, which may be defined as the resistance of a uniform column of mercury 106.3 cm. long and 1 square millimetre in section, at 0°.

The **unit current** is the **ampere**, which may be defined as a current which deposits silver at the rate of 0.001118 grams per second.

The **unit electromotive-force** is the **volt**, which is that E. M. F. that will cause a current of one ampere to flow in a circuit whose resistance is one ohm.

If **C** is the measure of a current in amperes, **R**, the resistance of the circuit in ohms, and **E**, the electromotive-force.

Ohm's Law may be expressed as follows:—

$$C = \frac{E}{R}.$$

QUESTIONS.

1. The electromotive-force of a battery is 10 volts, the resistance of the cells 10 ohms, and the resistance of the external circuit 20 ohms. What is the current ?

2. The difference in potential between a trolley wire and the rail is 500 volts. What current will flow through a conductor which joins them if the total resistance is 1,000 ohms ?

3. The potential-difference between the terminals of an incandescent lamp is 104 volts when one-half an ampere of current is passing through the filament. What is the resistance ?

4. A dynamo, the E. M. F. of which is 4 volts, is used for the purpose of copper-plating. If the resistance of the dynamo is $\frac{1}{100}$ of an ohm, what is the resistance of the bath and its connections when a current of 20 amperes is passing through it ?

5. What must be the E. M. F. of a battery to ring an electric bell which required a current of $\frac{1}{10}$ ampere, if the resistance of the bell and connection is 200 ohms, and the resistance of the battery 20 ohms ?

6. What must be the E. M. F. of a battery required to send a current of $\frac{1}{100}$ of an ampere through a telegraph line 100 miles long if the resistance of the wires is 10 ohms to the mile, the resistance of the instruments being 300 ohms, and of the battery 50 ohms, if the return current through the earth meets with no appreciable resistance ?

7. The potential-difference between the carbons of an arc lamp is 50 volts, and the resistance of the arc 2 ohms. If the arc exerts an opposing E. M. F. of its own of 30 volts, what is the current passing through the carbons ?

8. A dynamo, of which the E. M. F. is 3 volts, is used to decompose water. What is the total resistance in the circuit when a current of ½ ampere passes through it, if the counter electromotive-force of polarization of the electrodes is 1.5 volts?

9. On adding 3 ohms to the resistance of a certain circuit the current is diminished in the ratio of 6 to 5. What was the original resistance, and how much should be added to this in order to bring the current down to half its original value?

10. The current along a telegraph line is tested at two stations whose respective distances from the sending battery are 50 and 150 miles. The current in the latter case is one-half that in the former. If the galvanometer has a resistance equal to that of 15 miles of the line wire, prove that the battery resistance is equal to that of 35 miles of wire.

11. A dynamo gives an E. M. F. of 840 volts when running at the rate of 750 revolutions per minute, and its internal resistance is 12 ohms. Show that such a machine can supply 16 arc lamps in series, each lamp offering a resistance of 4.5 ohms, and requiring a current of 10 amperes.

---

2. Fall of Potential in a Circuit.

If a battery or dynamo is generating a current in a circuit, it is evident that the E.M.F. required to maintain this current in the whole circuit is greater than that required to overcome the resistance of a part of the circuit. For example, if the total resistance is 100 ohms, and the E. M. F. is 1000 volts, the current in the circuit is 10 amperes. Here an E.M.F. of 1000 volts is required to maintain a current of 10 amperes against a total resistance of 100 ohms, and it is manifest that to maintain this current in the part of the circuit of which the resistance is, say 50 or 75 ohms, an E.M.F. of but 500 or 750 volts will be required. This is usually expressed by saying that there is a fall in potential in the part of the circuit whose resistance is 50 ohms of 500 volts, and in the part of the circuit whose resistance is 75 ohms of 750 volts.

In general, if there is a closed circuit through which a current is flowing, **the fall in potential in any portion of the circuit is proportional to the resistance of that portion of the circuit.**

Experiment.

If your laboratory is supplied with a voltmeter, that is, a galvanometer wound and graduated to indicate directly in volts the difference in potential between two points of a circuit, measure the differences in potential between different parts of any circuit in which a current is flowing, and compare them with one another.

QUESTIONS.

1. The end, A, of the wire ABC is connected with the earth, and the difference in potential between the other end, C, and the earth is 100 volts. If the resistance of the portion AB is 9.6 ohms and that of BC 2.4, what current will flow along the wire, and what will be the potential-difference between the point B and the earth ?

2. The poles of a battery are connected by a wire 8 metres long, having a resistance of one-half an ohm per metre. If the E.M.F. of the battery is 7 volts and the internal resistance 10 ohms, find the distance between two points on the wire such that the potential-difference between them is 1 volt. What is the current in the wire ?

3. The potential-difference between the brushes of a dynamo supplying current to an incandescent lamp is 104 volts. If the resistance in the wires on the street leading from the dynamo to the house is 2 ohms, that of the wires in the house 2 ohms, and

that of the lamp 204 ohms, what is the fall in potential in the (1) wires on the street, (2) the wires in the house, and what is the potential-difference between the terminals of the lamp?

4. The potential-difference between the brushes of a dynamo supplying a current of 10 amperes to 38 arc lamps connected in series is 2000 volts. If the fall in potential in the connecting wires in the circuit is equal to the fall in two lamps, what is the fall in potential in a single lamp, and what the resistance in the connecting wires?

5. A cell has an internal resistance of 0.3 ohm, and its E.M.F. on open circuit is 1.8 volt. If the poles are connected by a conductor whose resistance is 1.2 ohm, what is the current produced, and what is the potential-difference between the poles of the cell?

6. The E.M.F. of a battery on open circuit is 15 volts. When the poles are connected by a copper wire a current of 1.5 amperes is produced, and the potential-difference between the battery poles falls to 9 volts. Find the resistance of the wire and the internal resistance of the battery.

II.—Measurement of Resistance.

3. Wheatstone Bridge.

The resistance of a conductor is usually measured by an instrument called a Wheatstone Bridge. It consists of a series of resistance coils made of German silver wire connected by conductors arranged in three sets A, B, and C, with connections for a battery, a galvanometer, and the resistance to be measured, as shown in Fig. 275.

The coils are mounted in a box, and the changes in the resistance are made by inserting or withdrawing conducting plugs, as shown in Fig. 276, or by turning switches as shown in Fig. 277

The branches A, and C usually have three coils each, the resistances of which are respectively 10, 100 and 1000 ohms, and the branch B has a combination of coils which

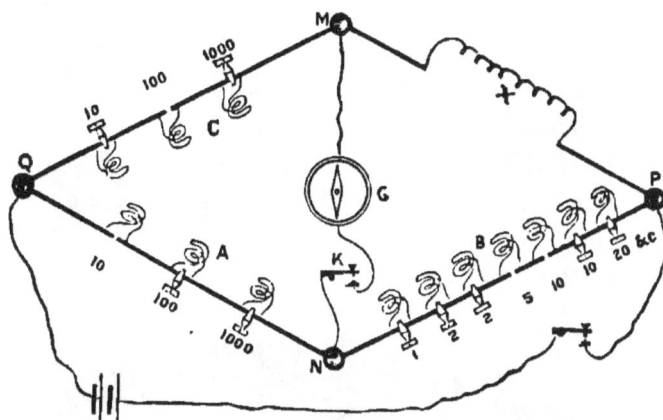

Fig. 275.

will give any number of units of resistance from 1 to 11,110 ohms. The conductor whose resistance (X) is to

Fig. 276.

be measured is inserted in the fourth branch of the bridge (Fig. 275), and the resistances A, B, and C adjusted until the galvanometer connecting M and N stands at zero when the keys are closed.

Then the current in the battery is flowing from P partly through X and C, and partly through B and A, to Q, and since no current flows from M to N, the potential of M must be the same as that of N, therefore the fall in potential from P to M in the circuit PMQ must equal the fall in potential from P to N in the circuit PNQ; but the fall in potential in a part of a circuit is proportional to the resistance of that portion of the circuit (Art. 2, page 411).

FIG. 277.

Hence

$$\frac{X}{C} = \frac{B}{A}$$

or

$$X = \frac{B \times C}{A}$$

The resistances A, B and C are read from the coils, and the calculations made.

Experiment 1.

Measure with a wheatstone bridge, the resistance of your galvanoscope, the coils used in Experiment 2, page 375, an incandescent lamp, etc.

Experiment 2.

Place two copper electrodes a distance apart in a solution of copper sulphate, and measure with the wheatstone bridge the resistance between them.

Would the result of the experiment be the same if platinum electrodes of the same size were used ? If not, why ?

4. Laws of Resistance.
Experiment 3.

Measure with the wheatstone bridge the resistances of pieces of the same wire whose lengths are proportional to 1, 2, 3, etc.

1. In what proportions are the resistances ?

2. What is the relation between the length of a wire and its resistance ?

Experiment 4.

Repeat Experiment 2, keeping the areas of the part of the electrodes immersed constant, and varying the distance between them.

How does an increase in the distance between the electrodes affect the resistance of the liquid conductor between them ?

Experiment 5.

Measure with screw calipers, or obtain from a table, the diameters of several copper wires of different sizes, and calculate the area of the cross-section of each. Measure the resistances of the wires.

What is the relation between the area of the cross-section of a wire and its resistance ?

Experiment 6.

Repeat Experiment 2; keeping the distance between the electrodes constant, and varying the areas of the parts immersed.

How does increasing the areas of the parts of the electrodes immersed affect the resistance of the liquid conductor between them ?

Experiment 7.

Obtain two wires, one of iron and one of copper, of the same size and length, and measure the resistance of each.

How many times as great is the resistance of the iron as the copper ?

The above experiments and others of a similar nature verify the following laws :—

Laws of Resistance.

1. **The resistance of a conductor varies directly as its length.**

2. **The resistance of a conductor varies inversely as the area of its cross-section. In a round conductor, therefore, the resistance varies inversely as the square of the diameter.**

3. **The resistance of a conductor of given length and cross-section depends upon the material of which it is made.**

5. Specific Resistance.

The resistance of a prism of length 1 cm. and cross-section 1 sq. cm. of any substance is the **resistivity,** or **specific resistance,** of that substance. It is generally estimated in microhms, or millionths of an ohm.

If l denotes the length of a conductor in centimetres, A the area of its cross-section in square centimetres, ρ its specific resistance, and R its resistance, by the laws of resistance

$$R = \frac{l\rho}{A}.$$

What will l and A represent in the case of a liquid conductor ?

6. Resistance and Temperature.

Experiment 8.

Place in the circuit with a battery a galvanoscope and a short piece of fine platinum wire. Observe the deflection of the needle of the galvanoscope, and heat the platinum wire with a spirit lamp or a Bunsen burner.

1. What change takes place in the current?
2. How can you account for it?

The resistance of nearly all pure metals increases about 0.4 per cent. for each increase in temperature of 1°C. The resistance of carbon diminishes on heating. The resistance of an electrolyte also decreases with increase in temperature.

———

QUESTIONS.

1. Copper wire one-twelfth of an inch in diameter has a resistance of 8 ohms per mile. What is the resistance of a mile of copper wire the diameter of which is $\frac{1}{36}$ in.?

2. If the resistance of a yard of iron wire, 0.03 inch in diameter, is 0.197 ohms, what is the resistance of 15 miles of iron wire, 0.3 inch in diameter?

3. What is the resistance of a column of mercury 2 metres long and 0.6 of a square millimetre in cross-section at 0°C.?

4. The resistance at 0° of a column of mercury 1 metre in length and 1 sq. mm. in cross-section is called a "Siemen's Unit." Find the value of this unit in terms of the ohm.

5. A mile of telegraph wire 2 mm. in diameter offers a resistance of 13 ohms. What is the resistance of 440 yards of wire 0.8 mm. in diameter made of the same material?

6. If the resistance at 0°C. of an iron wire 1 foot long and weighing 1 grain be 1.08 ohms, find the resistance of 0°C. of 1 mile of iron wire weighing 300 lbs.

27

7. What length of copper wire, having a diameter of 3 millimetres, has the same resistance as 10 metres of copper wire, having a diameter of 2 millimetres ?

8. On measuring the resistance of a piece of No. 30 B. W. G. (covered) copper wire, 18.12 yards long, I found it to have a resistance of 3.02 ohms. Another coil of the same wire had a resistance of 22.65 ohms ; what length of wire was there in the coil ?

9. The resistance of a bobbin of wire is measured and found to be 68 ohms ; a portion of the wire 2 metres in length is now cut off, and its resistance is found to be 0.75 ohms. What was the total length of wire on the bobbin ?

10. Two wires of the same length and material are found to have resistances of 4 and 9 ohms respectively. If the diameter of the first is 1 mm., what is the diameter of the second ?

11. What must be the thickness of copper wire which, taking equal lengths, gives the same resistance as iron wire 6.5 millimetres in diameter, the specific resistance of iron being six times that of copper ?

12. Two separate pounds of copper of the same quality are drawn out into uniform wires, the one twice the length of the other. Find the ratio between their resistances.

13. Two exactly equal pieces of copper are drawn into wire, one into a wire 10 feet long, and the other into a wire 20 feet long. If the resistance of the shorter wire is 0.5 ohm, what is the resistance of the longer wire ?

14. A piece of copper wire 100 yards long weighs 1 lb.; another piece of copper wire 500 yards long weighs ¼ lb. What are the relative resistances of the two wires?

15. Find the length of an iron wire one-twentieth of an inch in diameter which will have the same resistance as a copper wire one-sixtieth of an inch in diameter and 720 yards long, the conducting power of copper being six times that of iron.

16. A wire made of platinoid is found to have a resistance of 0.203 ohm per metre. The cross-section of the wire is 0.016 sq. cm. Express the specific resistance of platinoid in microhms.

17. Taking the specific resistance of copper as 1.642, calculate (1) the resistance of a kilometre of copper wire whose diameter is 1 millimetre ; (2) the resistance of a copper conductor one square centimetre in area of cross-section, and long enough to reach from Niagara to New York, reckoning this distance as 480 kilometres.

18. Two Grove cells, alike in all respects except that in one the plates are twice as far apart as in the other, are arranged in series, and the poles of the battery so constituted are united by a copper wire. The liquid in both cells becomes heated. In which is the rise in temperature the greater, and why ?

19. A current flows through a copper wire, which is thicker at one end than the other. If there is any difference either (1) in the strength of the current at, or (2) in the temperature of, the two ends of the wire, state how they differ from each other, and why.

III.—Grouping of Cells or Dynamos.

Cells or dynamos may be combined in various ways to give a current. The methods may be classified as follows :—

1. Series Arrangement.

7. Series Arrangement Defined.

Electrical generators are connected in series, or tandem, when the negative pole of the one is connected directly with the positive pole of the next (Fig. 278).

8. Effect of Series Arrangement on the Internal Resistance.

If n cells are arranged in series, and r is the internal resistance of each cell, it is evident that **the resistance of the group** = nr, because the current has to pass through a liquid conductor n times as long as that between the plates of a single cell.

9. Effect of Series Arrangement on E.M.F.

If the potential-difference between the plates of a single cell (Fig. 278) is e, the potential-difference between Z_1 and C_1 is e; but when C_1 and Z_2 are connected by a short thick conductor there is practically no fall in potential between them, therefore the potential-difference between Z_1 and Z_2 is e. Again, the potential-difference between Z_2 and C_2 is e, therefore the potential-difference between Z_1 and C_2 is $2e$. Similarly, for 3, 4, etc., cells the potential-differences are respectively $3e$, $4e$, etc. Hence the E.M.F. of n cell in series is $n\,e$.

Fig. 278.

10. The Current Given by Series Arrangement.

By Ohm's Law

$$C = \frac{E}{R}$$

but $E = n\,e$, and $R = n\,r + R_1$ where n is the number of cells, e the E.M.F. and r the internal resistance of each, and R_1 the external resistance in the circuit.

Hence

$$C = \frac{n\,e}{n\,r + R_1}.$$

2. Multiple Arrangement.

11. Multiple Arrangement Defined.

Generators are connected in multiple, or parallel, when all the positive poles are connected to one conductor, and all the negative poles to another, as shown in Fig. 279.

12. Effect of Multiple Arrangement on the Internal Resistance.

If n cells are arranged in multiple, and r is the internal resistance of a single cell,

the internal resistance of the group $= \dfrac{r}{n}$

because the current in passing through the liquid from one set of plates to the other has n paths opened up to it, and therefore the sectional area of the column of liquid traversed is n times that of one cell, hence the resistance is only $\dfrac{1}{n}$ of that of one cell. (Law 2, page 416.)

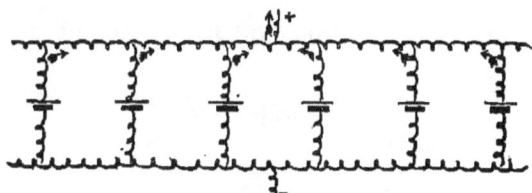

Fig. 279.

13. Effect of Multiple Arrangement on E.M.F.

When all the positive plates are connected they are of the same potential; for a similar reason all the negative plates are the same potential, hence **the E.M.F. of n cells in multiple is the same as that of one cell.**

This method of grouping has the effect of transforming a number of single cells into one large cell, the Z plates being united to form one large Z plate, and the C plates to form one large C plate. It must be remembered that **the potential-difference between the plates of a cell is independent of the size of the plates.**

Upon what is this potential-difference dependent?

14. The Current Given by Multiple Arrangement.

$$C = \frac{E}{R}$$

but $E = e$ and $R = \dfrac{r}{n} + R_1$,

where n is the number of cells, e the E. M. F. and r the internal resistance of each, and R_1 the external resistance.

Hence

$$C = \frac{e}{\dfrac{r}{n} + R_1}.$$

3. Multiple-Series Arrangement.

Sometimes both methods of arrangement are simul-

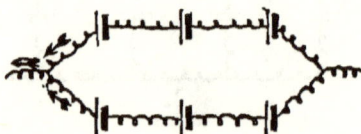

Fig. 280.

taneously employed, as shown in Figs. 280, 281.

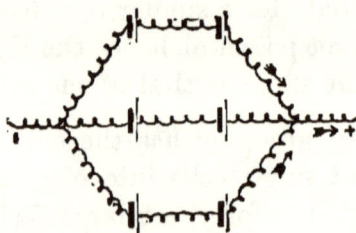

Fig. 281.

15. Current Given by Multiple-Series Arrangement.

To determine the current given by a number of cells grouped in multiple-series, consider each group in mul tiple as a single cell, and determine the current given by these groups when connected in series.

Thus, if n is the total number of cells, r is the internal resistance and e the E.M.F. of each cell, and m the number grouped in multiple,

the E.M.F. of each group in multiple $= e$

and the internal resistance $= \dfrac{r}{m}$

but there are $\dfrac{n}{m}$ such groups connected in series.

Hence

$$C = \frac{\dfrac{n}{m}e}{\dfrac{r}{m} \times \dfrac{n}{m} + R_1} = \frac{\dfrac{n}{m}e}{\dfrac{nr}{m^2} + R_1}$$

$$= \frac{ne}{\dfrac{nr}{m} + m\,R_1}$$

Where R_1 is the external resistance in the circuit. **The internal resistance of all the cells when grouped in this way is**

$$\frac{nr}{m^2}.$$

16. Best Arrangement of Cells.

It is manifest that when the external resistance is very great as compared with the internal resistance, to overcome the resistance, the electromotive-force must be increased, even at the expense of increasing the internal resistance, and the series arrangement of cells is the best. When the external resistance is very low as compared with the internal resistance, the object of the grouping is to lower as far as possible the internal resistance, and the multiple arrangement is the best. Between these extremes of high and low external resistance some form of multiple-series grouping gives the strongest current.

A general rule for determining the best method of grouping in any case may be found as follows:—

Let n denote the number of cells,

    r, the internal resistance of each cell,

    e, the E.M.F. of each cell,

    R_1, the external resistance,

    m, the number of cells in a multiple group when the current is a maximum.

Then

$$C = \frac{ne}{\dfrac{nr}{m} + mR_1} \qquad \text{(Art. 15.)}$$

$$= \frac{ne}{\left[\sqrt{\dfrac{nr}{m}} - \sqrt{mR_1}\,\right]^2 + 2\sqrt{nrR_1}}$$

C is a maximum when the denominator

$$\left[\sqrt{\dfrac{nr}{m}} - \sqrt{mR_1}\,\right]^2 + 2\sqrt{rnR_1}$$

is a minimum, but this quantity is a minimum when

$$\left[\sqrt{\dfrac{nr}{m}} - \sqrt{mR_1}\,\right]^2$$

is a minimum because $2\sqrt{2\,rnR_1}$ is constant, since r, n and R_1 remain always the same.

Now $\left[\sqrt{\dfrac{nr}{m}} - \sqrt{mR_1}\,\right]^2$ is a minimum when it equals zero, because a square cannot be less than zero.

Therefore the current is at a maximum when

$$\left[\sqrt{\frac{nr}{m}} - \sqrt{mR_1}\right]^2 = 0$$

or

$$\sqrt{\frac{nr}{m}} - \sqrt{mR_1} = 0$$

or

$$\frac{nr}{m} = mR_1$$

or

$$\frac{nr}{m^2} = R_1 .$$

but $\frac{nr}{m^2}$ is the internal resistance of the cells. (Art. 15.)

Hence

For a given external resistance the maximum current is obtained when the internal resistance is equal to the external resistance.

When

$$\frac{nr}{m^2} = R_1,$$

$$m = \sqrt{\frac{nr}{R_1}}.$$

Or

The current is a maximum when the cells are so arranged that the number in each group in multiple

$$= \sqrt{\frac{\text{internal resistance of one cell} \times \text{total number of cells}}{\text{external resistance}}} .$$

17. Example.

What is the best way of arranging a battery of 18 cells, each having a resistance of 1.8 ohm, so as to send the strongest current through an external resistance of 1 ohm ; and what is the current?

The number in each group in multiple

$$m \qquad = \sqrt{\frac{nr}{R}} = \sqrt{\frac{18 \times 1.8}{1}} = 5.69.$$

Since 6 is the factor of 18 nearest 5.69, the cells are to be arranged in 3 groups, in each of which 6 cells are joined in multiple, the groups being joined in series.

The E.M.F. of each group = 1 volt

and the internal resistance $= \dfrac{1.8}{6} = .3$

$$C = \frac{E}{R} = \frac{3 \times 1}{.3 \times 3 + 1} = 1.57 \text{ amperes.}$$

QUESTIONS.

1. 50 Grove's cells (E. M. F. of a Grove cell = 1.8 volt) are united in series, and the circuit is completed by a wire whose resistance is 15 ohms. Supposing the internal resistance of each cell to be 0.3 ohm, calculate the strength of the current.

2. Eight cells, each with half an ohm internal resistance, and 1.1 volts E. M. F., are connected (a) all in series, (b) all in parallel, (c) in two parallel sets of four cells each. Calculate the current sent in each case through a wire of resistance 0.8 ohm.

3. Ten voltaic cells, each of internal resistance 2 ohms and electro-motive-force 1.5 volts, are connected (a) in a single series, (b) in two series of five each, the like ends of the two series being joined together. If the terminals are in each case connected by a wire of resistance 10, show what is the strength of the current in the wire in each case.

4. You have 20 large Leclanché cells (E. M. F. = 1.5 volt, $r = 0.5$ ohm each) in a circuit in which the external resistance is 10 ohms. Find the strength of current which flows (a) when the cells are joined in simple series ; (b) all the zincs are united, and all the carbons united, in parallel arc ; (c) when the cells are arranged in groups of 2 in multiple ; (d) when the cells are arranged in groups of 4 in multiple.

5. The current from a battery of 4 equal cells is sent through a tangent galvanometer, the resistance of which, together with the

attached wires, is exactly equa₁ to that of a single cell. Show that the galvanometer deflection will be the same whether the cells are arranged all in multiple or all in series.

— 6. You are required to send a current of 2 amperes through an electro-magnet of 3.5 ohms resistance, and are supplied with a number of Grove cells each of 1.9 volt E. M. F., and 0.25 ohm internal resistance. How many cells are required ?

7. Calculate the number of cells required to produce a current of 50 milli-amperes (one one-thousandth of a₁ ampere), through a line 114 miles long, whose resistance is $12\frac{1}{2}$ ohms per mile, the available cells of the battery having each an internal resistance of 1.5 ohm, and an E. M. F. of 1.5 volt.

8. A current of not less than 0.016 ampere is to be sent through an external resistance of 360 ohms. What is the smallest number of Leclanché cells, each with E. M. F. 1.4 volts and resistance 15 ohms, by which this can be effected ? What would be the maximum strength of current obtainable if double this number of cells were used ?

9. The wire used on Indian telegraph lines is iron wire of No. 2 B.W.G., having a resistance of 4.6 ohms per mile. The batteries consist of cells of 1.04 volt E.M.F. and 30 ohms resistance per cell. Assuming that the resistance of the instruments is 80 ohms, and that a current of 8 milli-amperes is required to work them, find how many cells should be employed on a line 200 miles in length.

10. How would you arrange a battery of 12 cells, each of 0.6 ohm internal resistance, so as to send the strongest current through an electro-magnet of resistance 0.7 ohm ?

11. Find the best arrangements of 24 cells having an external resistance of 3 ohms, and each cell having an internal resistance of 2 ohms.

12. You have a battery of 48 Daniell cells, each of 6 ohms internal resistance, and are required to send the strongest possible current through an external resistance of 15 ohms : how would you group the cells ? Find also the current produced and the difference of potential between the poles of the battery, assuming that the E.M.F. of a Daniell cell is 1.07 volt.

13. You are supplied with 12 exactly similar cells, the internal resistance of each of which is one-fourth of the external resistance of the circuit: how would you group the cells so as to obtain the maximum current?

14. If you wish to heat a platinum wire, at a distance from the battery, to as high a temperature as possible, what sort of connecting wires will you use, and why? And what arrangement of battery-cells will you adopt?

If in the last case the insulation of the conducting wires was very imperfect, show whether it would be better to increase the number of cells arranged in series, or the number arranged in parallel; supposing that you have some additional cells at your disposal.

15. A circuit is formed of six similar cells in series and a wire of 10 ohms resistance. The E.M.F. of each cell is 1 volt and its resistance 5 ohm. Determine the difference of potential between the positive and negative poles of any one of the cells.

16. The internal resistance of a Daniell's cell is 1 ohm; its terminals are connected (a) by a wire whose resistance is 4 ohms, (b) by two wires in parallel, one of the wires having a resistance of 4 ohms, the resistance of the other wire being 1 ohm. Compare the currents through the cell in the two cases.

17. Six Daniell cells, for each of which $E=1.05$ volts, $r=0.5$ ohm, are joined in series. Three wires, X, Y, and Z, whose resistances are severally 3, 30, and 300 ohms, can be inserted between the poles of the battery. Determine the current which flows when each wire is inserted separately; also determine that which flows when they are all inserted at once in parallel.

18. The poles of a battery consisting of 40 Daniell cells in series are connected by a resistance of 280 ohms, and the current produced is 0.0535 ampere; when the external resistance is increased to 1080 ohms the current is reduced to one half: find the average resistance and E.M.F. of each cell of the battery, and determine the difference of potential existing between the poles of the battery when the external resistance is 280 ohms.

19. A Daniell cell, the internal resistance of which is 0.3 ohm, works through an external resistance of 1 ohm. What must be the resistance of another Daniell cell so that when it is joined up in series with the first and working through the same external resistance the current shall be the same as before? If the cells are joined up in parallel how will the current be modified?

ANSWERS.

EXERCISE I. Page 5.

1. $27\frac{3}{11}$; $3\frac{9}{22}$.
2. $48\frac{3}{5}$.
3. $2:1$; $11:6$.
4. $5:56$.
5. $\frac{1}{1200}$.
6. $\frac{44}{45}$.
7. 10.
8. (1) 400, (2) 800, (3) 200, (4) 200, (5) 800.
9. 10.
10. 0.98.

11. (1) 3520, (2) $58\frac{2}{3}$, (3) 7040, (4) $29\frac{1}{3}$.
12. 60 miles.
13. 7200.
14. $\frac{15}{22}\,ab$.
15. $36\,\dfrac{ch}{s}$ m.
16. (1) $\frac{5}{44}$, (2) $1\frac{3}{22}$.
17. $\frac{22}{15}\dfrac{bc}{a}$.

EXERCISE II. Page 8.

1. $4\frac{1}{2}$ ft. per sec.
2. (1) 10.5 cm. per sec., (2) 10 cm. per sec., (3) 11 cm. per sec.
3. (1) 1 cm. per sec., (2) 3 cm. per sec., (3) 1 cm. per sec., (4) 1 cm. per sec.
5. (1) 15 ft. per sec., (2) 75 ft.
6. (1) 35 cm. per sec., (2) 350 cm.
7. (1) 18 cm. per sec., (2) 14 cm. per sec., (3) 56 cm.
8. 175 ft. per sec.; 150 ft. per sec. ; 125 ft. per sec. ; 100 ft. per sec.; 540,000 ft.

9. 90 cm. per sec.
10. (1) 435 ft. per sec., (2) 44,250 ft., (3) 126,750 ft.
11. In 10 secs. from the instant its velocity was 20 ft. per sec. ; 1 ft.
12. (1) 4 secs. from the instant it was 8 ft. per sec., (2) 10 secs. from the instant it was 8 ft. per sec.
13. (1) 50 ft. per sec., (2) 250 ft.
14. 20 secs.; 800 cm.
15. (1) 5 cm. per sec. per sec., (2) 1750 cm.
16. $55\frac{5}{9}$.

EXERCISE III. Page 12.

1. 600 units of velocity ; 600.
2. 600 ft. per sec. ; 600.
3. (1) 300 cm. per sec., (2) 18,000 cm. per sec.

4. (1) $\frac{1}{5}$ ft. per sec., (2) $\frac{1}{300}$.
5. (1) 0.5 ft. per sec., (2) $\frac{1}{120}$.
6. 10 minutes.
7. 1 sec.

8. (1) 6, (2) 2, (3) $\frac{1}{10}$, (4) $\frac{1}{30}$.
9. (1) 50, (2) 5000, (3) $\frac{5}{6}$, (4) 83$\frac{1}{3}$.
10. (1) 1, (2) 3600, (3) 3600, (4) 60.
11. (1) 1200, (2) 72,000, (3) 720, (4) 12, (5) $\frac{1}{5}$.
12. (1) 30, (2) 108,000.

13. 38,400.
14. 35,280.
15. 2:1.
16. 1000.
17. 1000.
18. 1:2.
19. $\frac{1}{2}$.

EXERCISE IV.   Page 17.

1. 100 cm. per sec.
2. 20.
3. − 185 cm. per sec.
4. (1) 5, (2) 165 cm. per sec., (3) 20 sec. before its velocity was 100 cm. per sec.
5. (1) 10 sec., (2) 3$\frac{1}{8}$ sec.
6. (1) 550 cm., (2) 1 sec.
7. (1) 1.5 sec., (2) 11.25 cm. from starting point.
8. (1) 160 ft., (2) 250 ft., (3) 90 ft.
9. 156 ft.
10. 20 ft. per sec. per sec.
11. (1) 12, (2) 78 ft.
12. (1) 8 ft. per sec. per sec., (2) 10 ft. per sec., (3) 62 feet.
13. (1) 20 cm. per sec., (2) −5 cm. per sec. per sec.
14. 6 ft. per sec. per sec.

15. − 32 ft. per sec. per sec.
16. 2 sec. ; $\frac{7}{8}$ sec.
17. 16.12 sec. or 81.88 sec.
18. 40 ft. per sec.; 5 sec.
19. (1) 20 cm. per sec., (2) 4 sec.
20. (1) 20 ft. per sec. per sec., (2) 200 ft. per sec.
21. (1) 6 sec., (2) 144 cm.
22. 44 ft.
23. (1) 40 ft. per sec., (2) 35 ; 25 ; 15 ; 5 ft. per sec.
24. 2 : 1.
25. 23$\frac{1}{7}$ ft. per sec.
26. 20 cm. per sec. ; 20 cm. per sec. per sec.
27. Yes, if the body starts from rest.
28. 18 sec. from the time the first particle was at the given point ; 72 m. from the given point.

EXERCISE V.   Page 26.

1. (1) 160 ft. per sec., (2) 320 ft. per sec.
2. (1) 420 ft. per sec., (2) 260 ft. per sec.
3. (1) 1960 cm. per sec., (2) 980 cm. per sec.
4. (1) 256 ft., (2) 112 ft., (3) 156$\frac{1}{4}$ ft.
5. 49 m.
6. 156$\frac{1}{4}$ ft.
7. (1) 400 ft., (2) 16 ft.
8. 25 feet.

9. 100 m.
10. (1) 1$\frac{1}{2}$ sec. and 4$\frac{1}{2}$ sec., (2) 3 sec.
11. (1) 2$\frac{1}{2}$ sec., (2) 4$\frac{1}{2}$ sec.
12. (1) 6 sec., (2) 5 sec.
13. (1) 96 feet per sec., (2) 126 ft. per sec., (3) 80 ft. per sec.
14. (1) 36 ft. per sec., (2) 20 ft. per sec.
15. (1) 39$\frac{1}{16}$ ft., (2) 116.49 ft. per sec.

16. 7 m. per sec.
17. (1) 29.4 m. per sec., (2) 44.1 m. per sec.
18. 64 ft. per sec. ; 5 sec.
19. (1) 50 ft. per sec., (2) 150 ft.
20. 6 sec.; 176.4 m. from point of projection.
21. $6 \pm 3\sqrt{3}$ sec.
22. 12 sec.
23. (1) 204 ft., (2) 3 sec.
24. 1.547 sec.; 38.28 ft.
25. 4 sec. more.

26. In 1 sec. from instant of projection.
27. (1) 114 ft., (2) 144 ft.
28. $67\frac{1}{3}$ ft.
29. $52\frac{2}{3}$ ft.
30. 224 ft.
31. 160 ft.
32. 2 sec.
34. 1936 ft.
35. 1200 ft. per sec.
36. $\dfrac{b^2}{4a}$ ft.
37. $n \pm \sqrt{n^2 - 2n}$ sec.; $g\sqrt{n^2 - 2n}$.

EXERCISE VII. Page 36.

1. 35 gm.; 5 gm.
2. 2 P; 2 Q.
3. 39 pds.
4. 37 kgm.
5. 18 pds.
6. 12 P.
7. 12 pds. and 16 pds.

8. 20 gm. and 48 gm.
9. 6 pds. and $3\sqrt{5}$ pds.
10. 15 pds. and 20 pds.
11. 10 gm. and 24 gm.
12. 13 pds.
13. 15 pds.
14. 400 pds.

EXERCISE VIII. Page 39.

1. (1) 84 pds., (2) 18.477 pds., (3) 5.176 pds., (4) 70 pds., (5) 8.789 pds., (6) 2.125 pds., (7) 18.915 pds., (8) 12.64 pds., (9) P pds. north.
3. 17 pds.
4. 20 pds.
5. Forces are equal.
6. 8 gm.
7. 12 pds.
8. 12 pds. and 20 pds.
9. 10 pds. and $10\sqrt{2}$ pds.
10. 3 gm.; 1 gm.

11. 2 : 1.
12. $\sqrt{6}$ pds.
13. 50 pds. acting toward the centre.
14. $5\sqrt{2}$ kgm. at 135° with first force.
15. 5 : 4.
16. 10 and 26 pds.
17. $\frac{1}{2}\sqrt{7}$ times the side of the triangle.
26. Resultant is represented by OD = $2\sqrt{3}$ pds.
31. (1) $\sqrt{3}$ pds., (2) Resultant is represented by AD.

EXERCISE IX. Page 45.

1. (1) $5\sqrt{3}$ pds., (2) $5\sqrt{2}$ pds., (3) 2.58 pds.
2. $10\sqrt{3}$ and 10 pds.
3. $6\sqrt{2}$ pds.
4. $50\sqrt{2}$ pds.

5. $8\sqrt{3}$ and 8 pds.
6. $\frac{4}{3}\sqrt{3}$ pds.
7. 199.23 pds.
8. 98.48 pds.
11. (1) 6 pds., (2) 6 pds.

EXERCISE X. Page 49.

1. $3\sqrt{3}$ pds.
2. 22.2 or 10.56 pds. ; makes with the first force an angle whose tangent is $\frac{3}{4}\sqrt{3}$ or $\frac{1}{2}\sqrt{3}$.
3. 14 pds.
4. 28 pds. nearly.
5. $3\sqrt{3}$ pds. at 30° with third force.

7. 7.464 pds.
8. 5.477 pds.
9. 10 P toward opposite angular point.
10. Equals one of the forces in the direction of the sixth side.

EXERCISE XI. Page 54.

1. 40 and $20\sqrt{3}$ pds.
3. $1 : 1 : \sqrt{3}$.
4. 5 and $5\sqrt{3}$ pds.
5. 10 and 20 pds.
6. $2 : 1$.

7. $\dfrac{10\sqrt{2}}{1+\sqrt{3}}$ and $\dfrac{20}{1+\sqrt{3}}$ pds.

8. $\dfrac{2}{\sqrt{3}}$ pds.

9. $4\sqrt{3}$ pds.
10. $3\sqrt{2}$ pds.
11. 100 pds.
12. $10\sqrt{3}$ and 10 pds.
13. (1) $20\sqrt{3}$ pds., (2) 40 pds.
14. $\dfrac{100}{\sqrt{3}}$ pds. ; $\dfrac{100}{\sqrt{3}}$ or $\dfrac{200}{\sqrt{3}}$ pds.
15. (1) $\sqrt{3}$ pds., (2) 1 and $\sqrt{3}$ pds., (3) 45°.

EXERCISE XII Page 62.

4. 120° between any two forces.
6. (1) 120°, (2) 90°.
7. $1 : 1 : \sqrt{3}$.
8. 6 gm., $6\sqrt{3}$ gm.
9. $5\frac{1}{4}$ pds. ; $6\frac{3}{4}$ pds.
10. $7\frac{1}{2}$ pds. ; $12\frac{1}{2}$ pds.
11. $\dfrac{20}{\sqrt{3}}$ pds.
12. 38.4 pds. ; 28.8 pds.

13. 60 pds. ; 25 pds.
14. 74 pds.
15. $6\frac{2}{3}$ pds.
16. 6 pds.
17. 6 pds. ; 8 pds.
18. 5 pds. ; 13 pds.
19. 12 pds.
20. $\frac{1}{3}\sqrt{3}$ pds.
28. X lies in EH produced so that $AX = AE$.

EXERCISE XIII. Page 67.

1. 6 pds.
2. $10\sqrt{3}$ pds.
3. 5 gm.
4. 13 pds.
5. (1) 30 pds., (2) $30\sqrt{2}$ pds., (3) $30\sqrt{3}$ pds.

6. (1) 8 pds., (2) 32 pds.
7. 60 pds.
8. (1) $(50\sqrt{3}-5)$ pds., (2) $(50\sqrt{3}+5)$ pds.
9. (1) $2\frac{1}{2}$ pds., (2) $7\frac{1}{2}$ pds., (3) $(\frac{5}{2}\sqrt{15}-2\frac{1}{2})$ pds.

EXERCISE XIV. Page 76.

1. 2 pds.; 10.198 pds.
2. $\frac{1}{15}$.
3. 4.714.
4. $\sqrt{3}$.
5. $\sqrt{3}$; 1; $\dfrac{\sqrt{3}}{3}$.
6. $\dfrac{1}{\sqrt{3}}$.
7. 11.732 pds.

8. 36 pds.
9. 10 pds.
10. .732.
11. $24\frac{4}{9}$ pds.
12. $\frac{4}{3}\sqrt{3}$ pds.
13. .268.
14. 30°.

EXERCISE XV. Page 83.

1. (1) 4, (2) 576.
2. (1) 6, (2) 54.
3. (1) 50, (2) 500,000.
4. (1) 0.2, (2) 0.8.
5. 80 pds.
6. $\frac{9}{20}a$.
7. $30\frac{1}{4}$ pds.

8. 20 kgm.
9. $31\frac{1}{11}$ gm.
10. 1.5 sq. m.
11. $\frac{21}{60}$ in.
12. 35437.5 gm.
13. $1919\frac{3}{8}$ pds.
14. 7 kgm

EXERCISE XVI. Page 89.

1. 9.122 pds.
2. 0.0375 gm.
3. 11.5 pds.
4. 2.3 gm. per sq. mm.
5. 8:9.
6. 1:100.
7. 5:2:3.
8. 4:5.
9. 60 ft.

10. 9 m.
11. $31\frac{1}{4}$.
12. 9 kgm.
13. (1) $230\frac{2}{8}$ ft., (2) $1\frac{3}{8}$ ft.
14. 30 ft.
15. 1360.
16. 73.53 cm.
17. 4100.
18. 125 pds.

EXERCISE XVII. Page 95.

1. 5000 pds.
2. 30,000 kgm.
3. 3750 pds.
4. 107,500 kgm.
5. 77,000 pds.
6. 37,500 pds.
7. 72,000 kgm.
8. (1) 250,000 pds., (2) 62,500 pds., (3) 187,500 pds.
9. (1) 147,000 pds., (2) 48,-562$\frac{1}{2}$ pds., (3) 84,000 pds., (4) 63,000 pds.

10. (1) 42,000 pds., (2) 3600 pds., (3) 24,000 pds.
11. (1) 24,000 kgm., (2) 6000 kgm., (3) 10,500 kgm.
12. 182,000 kgm.
13. On each vertical face 187.5 pds. ; on upper face 156.25 pds. ; on lower face 218.75 pds.
14. (1) 212,500 pds., (2) 137,500 pds., (3) 87,500 pds.
15. 27 kgm.

16. 1500 pds.
17. 31.25 pds.
18. 100 ft.
19. 8 m.
20. 288 ft.
21. Upper side $2\frac{1}{3}$ in. below surface.
22. 7.25 ft.
23. 15 in.; top, 4.88 pds.; vertical side, 5.37 pds. ; bottom, 5.859 pds.
24. 2:1.
25. (1) 2.4 gm., (2) 0.64 gm., (3) 0.48 gm.
26. 2:1.
27. 4:13.
28. (1) 3080 kgm., (2) 8800 kgm.
29. (1) 11,160 gm., (2) 92,342$\frac{2}{7}$ gm.

30. (1) 9000 pds., (2) 2500 pd. . (3) 3000 pds.
31. (1) 905$\frac{1}{7}$ gm., (2) 1005$\frac{5}{7}$ gm., (3) 301$\frac{5}{7}$ gm.
32. 193,526 gm.
33. (1) 1848 kgm., (2) 792 kgm., (3) 1760 kgm.
34. 817$\frac{1}{7}$ kgm.
35. 8.
36. 18 in.
37. 400 gm.
38. (1) 416 kgm., (2) 1104 kgm., (3) 276 kgm., (4) 319.8 kgm., (5) 96.2 kgm.
39. 16:39.
40. (1) 14 in., (2) 64 in.
41. $2\sqrt{2}$ in. from the surface of the liquid.

EXERCISE XVIII.   Page 101.

1. 10 in.
2. 68 cm.; 170 cm.; 255 cm.
3. 2.427 ft.
4. 13.619.
5. 11:7.
6. 15 cm.
7. 14.96 cu. in.

8. $\frac{5}{242}$ of height of one arm.
9. Water 6.047 in. ; alcohol 5.953 in.
10. 4$\frac{1}{2}$ in.
11. At the bottom of the vertical tube containing the oil.

EXERCISE XIX.   Page 105.   -

1. (1) 62.5 pds., (2) 237.5 pds.
2. 4.96 pds.
3. 2.5 kgm.
4. 5 gm.
5. 295 pds.
6. 7 kgm.
7. 6.5 kgm.; 19993.5 kgm.
8. 968.75 pds.
9. 10 kgm.
10. $n+108$ pm.
11. 14,000 kgm.
12. 100 pds.

13. 2 c.dcm.
14. 6 cu. ft.
15. 20 cm.
16. 16 oz. ; 12 cu. in.
17. (1) 100.078 gm., (2) 60 c. cm.
18. 159.14 gm.
19. 311.9 gm.
20. 28.5 gm.
22. 1.5 gm. placed with the weights.
24. (1) 1$\frac{1}{2}$ c.cm., (2) 1 c.cm.

EXERCISE XX. Page 107.

1. 0.413 oz.
2. $\dfrac{n-m}{n}$ gm.
3. 0.881.
4. 5 c.dcm.
5. $4\frac{4}{5}$ cu. ft.
6. $\frac{1}{5}$.
7. 5.64.
8. $\frac{4}{5}$.
9. 225 pds.
10. n^2 c.cm.
11. $32\frac{1}{2}$ pds.
12. 520 gm.
13. 42 gm.
14. 9 pds.
15. 430 pds.
16. 41.56 in.
17. 216 sq. in.
— 18. 20 lbs.
19. 125,000 pds.

20. Forces are equal.
21. $40\frac{2}{5}$ lbs.
22. $13\frac{1}{5}$ lbs.
23. (1) 30 gm., (2) 20 gm.
24. 1000.
25. 3 cm. edge.
26. 1 : 6.
27. 10.65 gm. per c.cm.
28. 0.536 gm. per c.cm.
29. 4.75 in.
30. 23.48 cm.
31. 10,000 cu. ft.
32. 0.514.
33. 15 oz.
34. 0.64 in.
35. 30 lbs.
37. 1.458 pds.
38. (1) 2.604 pds., (2) 2.007 pds.
39. 7.102 pds. ; 50 pds.

EXERCISE XXI. Page 114.

1. 14.756 pds.
2. 1033.6 gm.
3. (1) 15 pds., (2) $14\frac{1}{2}$ pds., (3) 438.48 gm.
4. (1) 952 gm , (2) 1034 gm., (3) 1030 gm.
5. $7083\frac{1}{3}$ pds.
6. 24 ft.; $9\frac{3}{5}$ ft.; $7\frac{1}{5}$ ft.
7. 76 cm.; 152 cm.; 304 cm.
8. 272 in.; 240 in.; $81\frac{3}{5}$ in.
9. 1033.6 cm.
10. 1.43.
11. 1307.9 oz. per cu. ft.
12. 118.23 in.
13. $56\frac{2}{3}$ ft.

14. 12.5:1.
15. (a) 1, 22000 pds.; 8, (1) 1,100,000 pds., (2) 487,-500 pds., (3) 612,500 pds.; 13, $2,312\frac{1}{2}$ pds. on vertical face, $2,281\frac{1}{4}$ pds. on upper face, $2,343\frac{3}{4}$ pds. on lower face ; 18, 66 ft.

    (b) 2, 92,016 kgm., 28, (1) 18,997.44 kgm. ; (2) 99,756.8 kgm.
16. $2833\frac{1}{3}$.
17. 19.87 ft.

EXERCISE XXII. Page 118.

1. $(5)^5:(6)^5$.
2. (1) $(\frac{5}{6})^3$, (2) $(\frac{4}{5})^5$, (3) $(\frac{4}{5})^8$.
3. 3.
4. (1) $(\frac{3}{4})^5 \times 30$ in., (2) $(\frac{3}{4})^{15} \times 30$ in., (3) $(\frac{3}{4})^{20} \times 30$ in.

5. 4.
6. 11:1.
7. 3:1.
8. 75.12 gm.
10. 3:2.
11. 10.

EXERCISE XXIII. Page 124.

1. 10.336 m.
2. 17 ft.
3. $20\frac{5}{8}$ pds.
4. $114\frac{7}{12}$ pds.
5. $486\frac{1}{9}$ cm.
6. $43\frac{1}{5}$ sq. in.

EXERCISE XXIV. Page 127.

5. 34 ft.
6. 17 ft.
7. $h\frac{\rho}{\rho_1}$.

QUESTIONS. Page 152.

1. 1118.04 ft. per sec.
2. 4700 ft. per sec.
3. 6720 ft.
4. .952 sec.
9. $1109\frac{1}{3}$ ft. per sec.
10. 30.34 in.

QUESTIONS. Page 160.

8. 1 sec.; 3 sec.; 4 sec.

QUESTIONS. Page 164.

4. 52 in.; $1109\frac{1}{3}$ ft. per sec.
5. 33,792 cm. per sec.

QUESTIONS. Page 179.

1. 1, $\frac{1}{2}$, $\frac{1}{3}$, $\frac{1}{4}$, etc.
3. 1, $\frac{8}{9}$, $\frac{4}{5}$, $\frac{3}{4}$, etc.
5. W, $\frac{81}{64}$ W, $\frac{25}{16}$ W, etc.
6. 1, $\frac{1}{2}$, $\frac{1}{3}$, $\frac{1}{4}$, etc.
7. Equal.
8. 80 pds.
9. Divided by 2.
10. From C to G.
11. G of next higher octave.
12. Vibration-number will be halved.
13. $3\sqrt{10.5} : \sqrt{7.8}$.

QUESTIONS. Page 197.

1. $7\frac{1}{3}$ in.
2. 1126.4 ft. per sec.
3. 8.59 in.; 34.37 in.
5. 3:2.
6. 41.55 cm.; 124.65 cm.; 207.75 cm.; 83.1 cm.; 166.2 cm.; 249.3 cm.

QUESTIONS. Page 219.

1. 9:4.
2. 72:5.
3. 121:400.
5. 16.
6. 2 ft. from candle.

QUESTIONS. Page 409.

1. $\frac{1}{3}$ ampere.
2. $\frac{1}{2}$ ampere.
3. 208 ohms.
4. .19 ohm.
5. 22 volts.

6. 13.5 volts.
7. 10 amperes.
8. 3 ohms.
9. 15 ohms ; 15 ohms.

QUESTIONS. Page 411.

1. $8\frac{1}{2}$ amperes ; 80 volts.
2. 4 metres ; $\frac{1}{2}$ ampere.
3. 1 volt ; 1 volt ; 102 volts.

4. 50 volts ; 10 ohhs.
5. 1.2 amperes ; 1.44 volts.
6. 6 ohms ; 4 ohms.

QUESTIONS. Page 417.

1. 72 ohms.
2. 52.008 ohms.
3. 3.13 ohms.
4. 0.948 ohms.
5. 20.31 ohms.
6. 14.337 ohms.
7. 22.5 m.
8. 135.9 yds.
9. $181\frac{1}{3}$ m.

10. 0.6 mm.
11. 1.08 mm.
12. 4:1.
13. 2 ohms.
14. 1:100.
15. 1080 yds.
16. 3.04.
17. (1) 20.9 ohms, (2) 78.8 ohms.

QUESTIONS. Page 426.

1. 3 amperes.
2. $1\frac{5}{6}$ amperes ; $1\frac{10}{69}$ amperes ; $\frac{1}{5}$ amperes.
3. $\frac{1}{2}$ ampere in each case.
4. (a) 1.5 amperes, (b) 0.1496 ampere, (c) 1.2 amperes, (d) 0.702 ampere.
6. 5 cells.
7. 50 cells.
8. 5 cells in series ; 0.027 ampere.
9. 10.
10. 3 cells in each group in multiple.
11. 8 cells in each group in multiple.

12. 4 cells in each group in multiple ; 0.39 ampere ; 5.85 volts.
13. 8 cells in each group in multiple.
15. 0.75 volts.
16. 9 : 25.
17. Through X, 1.05 amperes ; through Y, 0.1909 ampere ; through Z, 0.0207 ampere ; through all three 1.105 amperes.
18. 1.07 volts ; 13 ohms ; 14.98 volts.
19. 1.3 ohms.

INDEX.